Wen-yuan Qian is currently a visiting scholar in the field of Chinese history and the history of science. He is attending a PhD programme at the University of Michigan and he will hold a three-year Distinguished Departmrent of History Fellowship.

CONTENTS

THE GREAT INERTIA

SCIENTIFIC STAGNATION IN TRADITIONAL CHINA

Wen-yuan Qian

科學停滯現象于皇權一統中國

巨大的慣性

錢文源

CROOM HELM
London ● Sydney ● Dover, New Hampshire

Croom Helm Ltd, Provident House, Burrell Row,
Beckenham, Kent BR3 1AT
Croom Helm Australia Pty Ltd, First Floor,
139 King Street, Sydney, NSW 2001, Australia

British Library Cataloguing in Publication Data

Qian, Wen-yuan
The great inertia.
1. Science—China—History
I. Title
509′.51 Q127.C5

ISBN 0-7099-2104-7

Croom Helm, 51 Washington Street, Dover,
New Hampshire 03820, USA

Library of Congress Cataloguing in Publication Data

Qian, Wen-yuan.
The great inertia.
Includes index.
1. Science—China. 2. Science—China—History.
3. Technology—China. 4. Technology—China—
History. I. Title.
Q127.C5Q24 1984 509.51 84-14217

ISBN 0-7099-2104-7

Photoset in Times by
Patrick and Anne Murphy Typesetters
Highcliffe, Dorset

Printed and bound in Great Britain by
Biddles Ltd, Guildford and King's Lynn

CONTENTS

this work is dedicated to
my beloved people
beyond the Pacific Ocean
among them are my
mother, wife, and daughter
who, owing to different reasons,
do not understand this work
but they do understand this worker
if not completely
certainly completely affectionately

fifteen years ago
it was this mutual love
that sustained me
from total collapse
in terror, torture,
in isolation, benightedness
and in extreme bitter feeling
and despair

now that I live
in a land
completely different
this love
though geographically remote
indeed infinitely profound
still supports me
day and night
in an arduous quest
for a better insight

PREFACE

Because of the author's unique abilities, this is a seminal book of the rare sort that ignites intellectual firestorms. We can see its significance only in the context of the great work by Joseph Needham that called it forth. But first we must put both the Needham Restoration and the Qian Reformation in historical perspective.

The flow of scientific technology outward from Europe has now penetrated all national cultures. The forces of technology and culture today buffet all peoples. Technology, being international, acts upon culture, which is national, to shape each people's modern way of life. In this process China has been laggard. The largest nation with the most persistent and distinctive culture has assimilated modern technology least effectively. It is still one of the poorest countries *per capita*.

The Needham Restoration has heightened this paradox by showing China's early technological superiority to Europe. The massive tomes of *Science and Civilization in China* have brought to light more and more of China's extraordinary achievements in observation of nature and development of material technology. Needham and his Chinese co-workers have restored to the Chinese people the record of their pioneer accomplishments in making chinaware and cast iron, paper and printed books, lock gates and the sternpost rudder, gunpowder, the compass and a host of other devices and inventions. Yet if China advanced so far so early, why did it fall behind in modern times? This paradox has haunted a whole generation of modern Chinese patriots and of historians of science.

The Qian Reformation launched in this book, *The Great Inertia*, attacks the paradox forthrightly in two ways, by asserting that (1) the Needham Restoration has in some respects claimed too much and yet (2) it has encompassed too little of China's history. The first assertion is made in Chapter 2 (Essay One) below on the level of specific phenomena, in Chapter 3 (Essay Two) on the level of historical factors and in Chapter 4 (Essay Three) on the level of philosophy of science.

The second assertion is made in Chapter 1 (an Introduction), which argues that China's 'intolerant politico-ideological environment' inhibited the establishment of science as an autonomous social institution; and the resulting domination of thought by political authority, a purely negative factor, is not given proper weight in the Needham Restoration. In other words, Joseph Needham's magnificent feat of surveying and discovering the Chinese record in the various disciplines of science has understandably stopped short of encompassing the whole of Chinese society. The Needham volumes deal with the technology of government only in passing, yet many historians now see the all-powerful Chinese state as the greatest of all China's technological-social achievements. For example, the Legalist technique of popular self-policing by mutual surveillance under the doctrine of collective responsibility, like the Confucian arrangement for each scholar's self-indoctrination in the classical teaching of social order through his arduous preparation for civil service examinations, are only two of the Chinese inventions in political technology. In sum, they perfected government by bureaucracy.

The early Chinese success in the organisation of their state and society all across the board — economically, politically, socially and culturally — accompanied the efflorescence of material technology and invention that Dr Needham and his colleagues justly celebrate. All these achievements came along together. But in the process the growth of science as a social institution was smothered by the predominant growth of the state power. Chinese creativity in social and political engineering rivalled their inventiveness in material technology, but as the great institutions of autocracy and bureaucracy became deeply established, they stifled creativity. Modern revolutionists finally had to attack the Eight Legged Essay of the Ming-Ch'ing examination system as hobbling men's minds just as clearly as footbinding hobbled Chinese women. China's early prowess in organisation had strangled her capacity for growth.

This I think is the truth to which Qian Wenyuan's Reformation leads us. As a Chinese intellectual of the Maoist era, he sees China's tardiness in the assimilation of modern science as an aspect of the whole social process. He lacks both the sentimentality of foreign benefactors and the zeal of latter-day missionaries-in-reverse. He is conscious mainly of his patriotism and of the scholar's function to seek the truth. The power of the Chinese state has been made all too evident to him.

Wen-yuan Qian (Ch'ien) was born in 1936 and in 1959 finished a five-year Soviet-type training at Peking University in theoretical physics. He then spent 21 years teaching at Zhejiang (Chekiang) University in Hangchow. There he taught theoretical physics from 1959 to 1966 and after the Cultural Revolution from 1977 to 1980, whereupon he came to Northwestern University. There, after a short time in the Physics Department, he transferred to the History Department and in 1983 received his MA in history. He now holds a three-year Distinguished Department of History Fellowship for PhD work at the University of Michigan, Ann Arbor.

During the long decade of the Cultural Revolution 1966–76, Qian tells us that 'at the cost of extreme personal agony, I learned what communist totalitarianism means', yet he says he did not waste his time and was able to develop his command of English. An outsider untouched by the revolution can only marvel at this man's accomplishment in making his way from China to America and intellectually from physics to the history of science. On the one occasion when I met Mr Qian he seemed to me thoroughly objective and remarkably unaggressive, yet his mind has been prepared by long academic training and by intensely personal experience to deal succinctly with large issues of history. The reader will soon see that this Preface has by no means encompassed the range of Qian's thinking, any more than I can comprehend all the dazzling detail of Joseph Needham's masterwork. My suggestion that in the study of China's history the state power is the frontier most in need of exploration, is only one angle of approach. William Monter's Foreword puts it in a broader context.

John King Fairbank

FOREWORD

Readers need to know three things about the author of *The Great Inertia*. Wen-yuan Qian is a physicist who admires the taut beauty of his discipline; he is a deeply patriotic Chinese; and he possesses a stubbornly independent turn of mind ('incurably bourgeois', in the Maoist idiom). Every sentence of his work reflects at least one of these aspects, and many combine all three. Only a man endowed with unusually tenacious idealism would undertake a major enterprise in a new discipline in middle age, written in a foreign language; only someone with unusual originality of mind could have given it the shape he has. The result is a serious work of comparative macro-history, an overview of the non-development of science in China seen against the background of European science, and occasionally Japanese technology.

Across its path lies the voluminous work of Joseph Needham, the principal expositor of traditional Chinese science to the larger audience of English-reading non-experts — to which I belong. Qian's perspectives, both as a physicist and as a Chinese, inform a critique which challenges some of Needham's fundamental assumptions and thus becomes the prolegomenon to a different history of traditional Chinese science. If you begin with the physical rather than the biological branches of science, and if your point of departure is the present-day 'neo-colonial' or 'third-world' state of Chinese science instead of the technological precocity of China a thousand or more years ago, then Qian rather than Needham provides the better historical guide for the neophyte.

But, as a culture-bound European historian, my principal concern with Qian's work lies in its assumptions about, and implications for, the history of European science. If it remains true that the only world-historical scientific revolution occurred in Europe about three centuries ago, *and* if Qian is correct in his assumption that 'software decides', then exactly what types of non-scientific 'software' accompanied, facilitated, or at any rate failed to prevent the emergence of a recognisably 'modern' science by the time of Newton?

Cultural pluralism, says Qian. Considering that the human mass of China and Europe (including western Russia) approximately balanced each other, the most important contrast he sees is between the politically and culturally unified Chinese empire, held together by a Confucian ideology administered by a bureaucracy equipped with a superb rote memory of classical texts — and the multiple divisions of Europe, where the loyalties of intellectuals were split between church and state, between state and state, and (after Luther) between church and church. In China a capricious, free-will 'Son of Heaven' presided over an irregular natural world, but human affairs were usually well-ordered; in Christendom, a watch-maker-God regulated Nature while human affairs swirled in a disorderly cauldron.

Qian's macro-historical comparison raises important questions about the origins of modern science in Europe. Thus far, the best known investigations into the cultural context of early modern science have tended to concentrate on the religious affiliations of important scientists — the 'Puritanism and Science' problem first raised by R. K. Merton over 50 years ago. But Qian's arguments suggest that we should shift our focus slightly, placing Europe's early scientists not so much between conflicting confessions as between secular and religious values, or plunging more deeply into the problem of the language of scientific communication within a divided Christendom. Certainly we historians of Europe are aware of, indeed we celebrate, the variety of national cultures that flourished in Europe long before the rise of modern science. Yet we also recognise that science is 'naturally' an international phenomenon, developed by scattered individual talents who must maintain frequent communication with one another.

Viewed from these angles, the early history of modern science in Europe seems full of riddles. Consider the basic 'software' of language: the 1600s, the time when Qian and Needham agree that Europe had definitively surpassed the best achievements of Chinese science, also marked the moment when Latin ceased to be the basic medium of communication among Europe's 'literate' people. Galileo was the first major scientist to abandon Latin; Newton's *Principia* was the last important book to be composed in Latin. Latin remained the language of the universities and of the Popes; but seventeenth-century science was notoriously something that matured outside the universities (both Galileo and Newton abandoned their chairs of mathematics as soon as fame made it possible to do so, claiming to better devote themselves to scientific pursuits), and despite the Popes (again Galileo provides an all-too-famous illustration). Latin became a 'dead' language — but it continued to be learned by Europe's mandarins right down to the age when Joseph Needham attended its elitist universities. There was less difference between European universities and the imperial Chinese civil service than Qian wishes to admit.

Another set of problems grow out of Galileo's struggle with the Roman Inquisition, and the subsequent founding of royal academies for the advancement of science by the leading monarchies of Western Europe, England and France, in the 1660s. If Qian is correct in pointing to the division of loyalties between church and state as one of Europe's particular strengths, then what happened at the crucial moment of birth was simply that the state replaced the church as the successful patron of modern science. But the state was a less logical patron than the church, because the church employed an international language of communication while national states practised an inherent linguistic chauvinism; moreover, states were inherently more suspicious of an international enterprise than was a church whose interests had always been international. Perhaps the condemnation of Galileo was a worse tragedy than we recognise.

Did European science emerge between 1600 and 1660 because of, or despite, this cultural and religious pluralism? This age witnessed the zenith of confessional conflict in the Thirty Years War; it executed more witches than at any other time in history; it suffered from serious attacks of bubonic plague, still incurable after three centuries of familiarity; and it was even scarred by a sustained international economic depression. Yet by 1640 Europe had surpassed China's greatest achievements in the area which Needham considers their finest — magnetism — and equalled China in that field where Qian sees their greatest achievement, a mathematical theory of the musical scale. Perhaps Qian has correctly identified the tensions between church(es) and state(s) as providing enough opportunities for a 'Republic of Letters' to incubate and to disseminate early modern science, despite all attempts to censor it (Urban VIII) or to nationalise it.

Galileo was published (in Latin) by Protestant Europe precisely because he had been condemned by the Pope. But this does not mean that liberty of thought existed in seventeenth-century Europe. Granted, there were eloquent spokesmen for free thought; one thinks of Milton's *Areopagitica*, or of the even more daring arguments in Spinoza's *Theological-Political Treatise* of 1670. Yet these men could write 'freely' only because of their proximity to powerful statesmen who protected them from censorship. Milton's patron, Cromwell, enforced an unprecedented degree of religious toleration in a major state because he was a military dictator (there is a spectacle for Confucius and Mao to wonder at!)

In this work, Qian offers us a provocative approach, which among its other merits calls attention to the full range of cultural dimensions within which the history of Chinese, and European, science lies embedded. His stance is indeed 'incurably bourgeois', resembling that of nineteenth-century European liberals. But let us not forget that nineteenth-century European liberalism had an excellent effect on the study of history. At his best, Qian resembles one of the giants among nineteenth-century liberal historians — a deeply 'patriotic' Catholic with a passionate belief in historical impartiality and in the possibility of human progress through self-critical historical understanding. Qian tells me that he has never read Lord Acton nor heard of his dictum that 'the permanent and universal question is whether the living shall govern the dead', that we should study the past in order to overcome it. None the less, that is exactly what he tries to do here.

William Monter
Northwestern University

THE GREAT INERTIA

EXPLANATIONS AND ACKNOWLEDGEMENTS[1]

> Man is born for uprightness.
>
> — Confucius, *Lun Yü*

> All men in some degree need outside approval for what they think and do — the approval of loved ones, of kith and kin, of the community of right-minded men. Most men in all times obtain the required approval by following the established customs and professing the common opinions. But there are always some eccentric individuals, and on occasion certain groups, who find the present temporal world of men and things intolerable. So they withdraw from it, living in spiritual exile, or else they endeavor to transform it.
>
> — C. Becker, *The Heavenly City*[2]

According to Carl Becker, I am doubtless an eccentric individual, both for my views on China's present general condition and the prevailing historical interpretation of China's scientific past. But to hold those views of mine has been my harrowing experience. From time to time I need the above-quoted pithy Confucion adage to brace myself.

I think we present-day Chinese should be proud of being Chinese, because we are challenged by a double task: that of coming to terms with our long glorious heritage while confronting the tremendous mission of creating a future different from any previously envisioned. The formidable burdens of the past and of the present require a patriotic and thinking Chinese to explore ways of meeting this historic task. China's paradoxical delay and its distressing present-day experience should not merely embitter us, but should also enable us to become enlightened by truth, which in its turn might be useful to other peoples in need of democratisation. We are suffering; we are also lucky — because suffering should oblige us to be concerned with deeper questions and to think more.

For more than 20 years I served as a university teacher of

theoretical physics and as a Maoist 'negative teacher' (fan-mian jiao-yuan), a 'rightist' or an 'ideological counter-revolutionary' who lives among and under the 'supervision' of 'the masses', that is, under the constant harassments and persecution of the Party (the CCP — Chinese Communist Party). Throughout this time I had been aware that my proper study is my disaster-ridden China. But I was not clear about what I could personally contribute to the diagnosis of China's chronic illness, and to the relief of China's traditional and endless misfortune. What Dr W. E. B. Du Bois has said about Africa in his *Autobiography* has elicited in my heart a strong feeling:

> Africa, all I ask from you is the courage to know: to look about you and see what is happening in this old and tired world; to realize the extent and depth of its rebirth and the promise which glows on your hills.

I have often had the same thoughts about my China.

But poetic empathy did not serve as a specific guide. Several years ago, I knew only that I was by nature a thoughtful scholar; I had been trained and was working on theoretical physics; among my Chinese peers, my English was exceptionally good; and I was very much interested in history. To be fair, it was not just academic interest. I acquired these abilities all through purposive and arduous labour. My parents did not speak a single complete sentence of English, and historical scholarship is not often entertaining reading. My effort may be summarised as a quest for historical insight and a comparative perspective, and my aim, as indicated before, is a causal diagnosis of my nation's long-standing basic problems.

In retrospect I realise that the combination of my interests, or abilities, offers me an opportunity to serve my mission in a peculiar — yet also significant — way. Theoretically, the opportunity is based on an 'axiom': a nation's scientific stagnation is an inseparable component of its general historic-social stagnation. More specifically, we may hypothesise: scientific stagnation is historically a necessary condition of general stagnation, though not necessarily a sufficient condition. There is no doubt that it requires further supportive arguments, but I believe that many people share my confidence in this 'axiom'. This plausible intimate relationship between general development and scientific development enables me to inquire about China's general stagnation by way of China's scientific stagnation. Besides, my increasing command of the English language enables me to enter the stage of cosmopolitan academic debate, no longer fettered by the CCP's persistent bigotry. It was not the allure of international publicity, but, owing to the current Chinese situation, the much simpler desire for seeking truth that urged me to overcome the language barrier. Finally, on a more practical note, I am offered the opportunity to contribute to the debate over China's scientific past, which has already encouraged considerable research. Several famous Western scholars have not only described the results of their scholarship, but have also expounded a series of theoretical explanations — often, I would say, lacking a logical basis, but with an authoritative air. Fot instance, a hypothetical and evidently wrong order of physical discoveries was proposed on the basis of an overzealous appraisal of the history of Chinese physics. Understandably, I found that, as a Chinese physicist now doing history, I was duty-bound to pick up the gauntlet. By taking issue with theories such as this, step by step, an independent work emerged.

Three years ago, I had not thought of finishing this particular study or of composing a book-size work in this language. That I wound up studying in a free political environment, at a top-level institute, and that an old dog still learns new tricks — well enough to finish this work to my own and others' satisfaction — such developments were 'historically indeterminate', that is, fortuitous. Who could guess that Maoist China would open to the West since the end of the 1970s in a manner unprecedented in the history of communism? In fact, the very title of the work cropped up unexpectedly when I was asked to speak about 'Science and Technology in Present-day China'. My thinking in the past three years has moved from contemporary China to traditional China, from traditional China to ancient feudal China, from phenomenological description to causal questioning, from the non-development of modern science to the general inhibition of 'modernity', from Sino-European comparisons to other contrasts, from the common features of various comparative cases to a general theory of 'modernisation', and from historical studies to the understanding of the crucial problems of present-day China (that is, from the analysis of oriental despotism to the denunciation of communist totalitarianism). This study started accidentally, yet for me, the course of its development was in fact 'historically determined'.

So, intellectually I have made a round trip. It is now clear to me that for the rest of my life I am destined to make this round trip again and again and again until one day I fall *en route* for whatever reason.

Another unexpected circumstance is that this work of mine, which may be called my 'anti-Chinese-interia', would finally embrace such a sweeping yet completely natural amalgam of historical analysis, scholarly criticism, philosophical argumentation, and politico-ideological polemics. First I thought I was writing an essay, debating with other essays on some logical issues, in order to justify the idea of Chinese inertia. As the work grew, it also branched. As it branched, I believe, it also acquired the capability to stand more firmly and independently.

Some technical explanations about the work are in order. This is a collection of four articles, each of which was written somewhat independently, and I have kept this relative independence. When cited in this collection the four articles will be referred to, in their present order, as the Introduction, Essay One, Essay Two, and Essay Three (Chapters 1, 2, 3 and 4). Since I became aware of the possibility of putting these articles together, with the help of my advisors and my young friends, I have tried to keep every useful idea, while avoiding repetition, and at the same time have set out the material and arguments in a more organised form.

All these chapters contain a number of extended notes. I hope that my readers treat these notes as seriously as they treat the text.

All Chinese names, except for very familiar ones like Confucius and Mencius, are spelled with the new system of Romanisation, 'pin-yin'. This rule covers all quoted passages, except when I could not guess the original Han characters. I use the 'pin-yin' system simply because I am familiar with it, and because my name is now officially spelled as it appears here; I know that 'Qian' looks and hence sounds strange to native speakers of English. I think, historically speaking, that this system (which is good, but with a few exceptions, notably q, x, and c) reminds us of the fact that it was designed when mainland China was very much separated from the rest of the world.

Although in this work I have proposed two 'axioms' — 'software' decides and 'degree of activation' counts — I do not mean to propound 'axiomatisation' generally in socio-historical studies, as theoretical physicists and mathematicians do in their fields. By 'axiomatisation' I mean unified understanding on the basis of comprehensive principles — a small bundle of axioms, not necessarily self evident, but very fertile in producing theoretical derivatives and very sturdy in withstanding empirical and logical tests. In the realm of social studies, all post-Newton thinkers were influenced in varying degrees by Newton's (or Descartes', or Euclid's) axiomatic methodology. Jeremy Bentham proposed his Utilitarian Principle as *the* correct basis for understanding and guiding social phenomena; Adam Smith summarised the division of labour, *laissez faire*, etc. If Newton and Euclid's axioms have been strictly followed, Bentham and Smith's principles have proved to be enlightening to their followers, though not strictly applied; their theoretical and aesthetic appeal is far superior to their direct applications. The differences between physical and social phenomena are just too big; the former are characterised by mathematical inexorability (including quantum-mechanical probabilistic laws), whereas the latter are too much mixed with and influenced by human free will, natural environment, cultural history, and current conditions. On the other hand, the axiomatic line of thinking can help us identify some macro-historical, hence significant, causes and motifs.

For over a century China has been in a 'translation movement'. This is literally true in scientific and cultural matters, and to a considerable extent in various forms of official ideology, whose major representatives are Hong Xiu-quan, Sun Yat-sen, Mao Ze-dong, and Deng Xiao-ping. History witnesses that a trend of national optimism that existed among revolutionary intellectuals in the first few decades of this century was illusory. During that period of idealism and action, many writers expressed the view that once a social revolution took place, the ancient Chinese civilisation would quickly reassert itself in a remoulded form, and China would go ahead by leaps and bounds. This overly sanguine point of view did not recognise the Chinese inertia. More than 60 years' tortuous actual development refutes this view. Logically, it is also very understandable that present-day Chinese readers are much surprised when they have recently been told by an American professor that a 'scientific revolution' had occurred in seventeenth-century China. But on this issue I think we need to emphasise a 'Tolstoyan principle of historical commensurateness.'[3] In Tolstoy's

War and Peace three peasants argued about what makes a locomotive move. Metaphorically their futile argument illustrates an important principle — in Tolstoy's own formulation:

> The only conception capable of explaining the movement of the locomotive is that of a force commensurate with the movement observed. The only conception capable of explaining the movement of peoples is that of some force commensurate with the whole movement of the peoples.
>
> Yet to supply this conception various historians assume forces of entirely different kinds, all of which are incommensurate with the movement observed.

A historian who neglects this Tolstoyan principle is analogous to a physicist who forgets the law of energy conservation.

According to my understanding, the word 'revolution' is generally reserved for historic-social changes of gigantic scale and momentous consequences — for instance, the 'price revolution' in sixteenth-century Europe. Therefore, I would use 'revolution' to describe such events and try to correlate them with commensurate causes and effects. Although since the 1960s 'scientific revolutions' have proliferated in historiography, I think that most of them can better be called scientific crises that led to breakthroughs.

In any case, there has never been a 'scientific revolution' in China, past or present. Besides, so far the concept of an 'ecumenical science' has never historically conformed to reality. I would add that the ongoing Chinese 'translation movement' differs remarkably from the one that occurred in medieval Europe, when European schoolmen translated proto-sciences and other classics, and when Western creativity surpassed the ancient originals and forged ahead within a few centuries (the fourteenth-century was a scientific peak). If China — with about one-fourth of the world's population — wants to catch up with the West and forge ahead, it is vital to pay *foremost* (indeed for many Chinese *painful*) attention to 'software' reformation, thus to raise 'the degree of intellectual activation' in a genuine, all-round, and enduring manner.

The work contains some personal observations (notes 5 and 14 of Essay Two (Chapter 3), Section 16 of Essay One (Chapter 2), etc.) outside this preface. Those passages occurred so naturally and spontaneously that I prefer to let them stay where they are. Besides, the material in Section 16 of Essay One is actually not personal. It is the phenomenology of Chinese history — the continuation of medieval Chinese history, as well as modern Chinese history — that is the Chinese inertia. (Both my grandmothers, one from a very rich family and one from a very poor family, had bound feet.) Another reason is that I prefer that this work be understood as synthetic in nature, because that implies a unified theoretical grasp — either 'vertical' (the past and the present), or 'horizontal' (different aspects of social life at any given time). Some of my readers will consider those notes and some other passages as political polemics. They will be correct, although for the sake of 'academic purity', I have tried hard to limit ideological preaching and denunciation to a minimum, including only the most obvious inferences and strictly logical extrapolations. I was brought up with Marxist indoctrination, and some influences are still with me. Among others I like the idea that scholarly pursuits can be combined with a revolutionary passion.

If confessing my interest in political polemics leaves an impression on some readers that I am merely doing pamphleteering, shaping scholarly arguments to serve a right-wing political purpose, then I would think it a most sad misconception about me. My true logic — that is, my personality — is that if I cannot put my 'right-wing' political faith in intellectually convincing (as personally honest and comprehensive as possible) arguments, then I am going to surrender myself to Beijing tomorrow morning. If my refutations of some scholarly hypotheses and arguments are disclosed tomorrow morning as incorrect or untenable, then I will make self-criticisms to concerned scholars tomorrow afternoon. (Such statements sound very un-American, I know. But I do not care about that. Every society at any time has its mainstream or dominant sets of values, sometimes vulgar, deceptive, or cruel. And every society has its non-conformists. When in China, my voice sounded very 'un-Chinese', and because of that, I was threatened with execution squads or long-term imprisonment — in Maoist China that means a lifelong concentration camp sentence. Then I learned to lie, and I have lied enough! Being un-American a person can still live in America, and can even find an understanding audience — that is the very beauty of America!) It was in fact my learned conclusion that China can only be understood in a manner that unifies traditional China and present-day China (cf. note 7). that precipitated my personal political crisis: I knew that I could not go back; from now on I would have even more difficulty to restrain myself from telling people what I believe to be true and to mask my contempt towards hypocritical despots; and thus going back means an even harsher treatment to my family than being geographically separated from them.

After reading parts of my writing, a young colleague of mine said: 'You have the inclination to tell your readers where your ideas were from. "One beautiful sunny morning, this bright idea struck me. . ."' Very sharp observation indeed. I first explained it as my historical propensity. Then I understood that if this explanation is correct, it is only partly correct. The fact is that I lived under the rule of Mao for more than 25 years. Being so stubborn a 'bourgeois intellectual', in those years I was always in a state of vigilance. Month after month, year after year, I had to be prepared to undertake 'self-criticisms' at any time, and everyone of us intellectuals was asked to be absolutely honest to the Party. Isn't it ironic and unfair that an absolutely hypocritical authority asks its subjects to be absolutely honest to it? Absolutely honest or not, anyway, I must have formed this habit of detailing the source of an idea in the long course of continuously handing over to the Party leadership my confessional reports.

I know that in recounting the above episode and reminiscence, I am again acting in accordance with this old habit of mine. I can't help it.

As a neophyte in historical research and as a foreigner writing in English, I am extremely fortunate to have benefited enormously from many scholarly friends, from distinguished academics to undergraduates, who helped me with my research, my writing, my formal coursework, and my linguistic training. Among them I want, first of all, to mention proudly my mentors, Professor David Joravsky and Professor William Monter. Since this is the work of a man who had spent 44 years listening to, talking, reading, and writing Chinese (mixed with a small but basic proportion of English), including 28 years in three departments of physics (either as a student, a teacher, or a visiting scholar, with a small but effective proportion of time devoted to historical studies), my readers will easily believe that this work became possible *only* because of the effective guidance and help of my eminent mentors and good friends. Moreover, I must add that their help was characterised by complete sincerity, full understanding, and a critical attitude. Other professors of the Department of History at Northwestern University James Sheridan, Thomas William Heyck, Conrad Totman, and Betty Jo Dobbs also gave me valuable assistance. During my years at Northwestern, I have always first discussed my ideas on the history of science with Dr Robert Michaelson, the Head Librarian of Northwestern's Science and Engineering Library. Over a period of several months, my visiting him at his office or calling him by phone became my daily routine. His contribution to this work and sympathy for my situation are memorable.

For two years, a letter from Professor John K. Fairbank has inspired me constantly. The letter begins with this sentence: 'Your Introduction (sent me July 23, 1981) makes very exciting reading because we badly need to bring new vitality into historical studies by attacking the big questions as you are doing.'

At an early stage of this work, I had an unforgettable opportunity of explaining my themes to Professor Freeman Dyson, an outstanding theoretical physicist, indeed one of the best of this century. His enthusiasm for my novel ideas and his request to read my work often spurred me when I felt tired over the unwieldy typewriter.

Now, when I believe I have finished some meaningful work, I feel thrilled each time I recall the moral support that right-minded people imparted to me, the long discussions I had with my American friends, and the warm welcome my talks on related topics received on various occasions. To me, a man who recently moved from an allegedly socialist nation to a presumably individualist society, all this assistance strikes me as an affectionate and unexpected gift. The work is truly a brainchild of collective wisdom and teamwork. Yet, there is no question that I am the only person responsible for its particular stand, basic ideas, and stylistic drawbacks.

Among many perceptive and generous friends who helped me to become an historian using English, I must also mention Professor Laurie Brown (historian of physics), Professor Richard Lerner (Truman College), Professor Sarah Maza, Professor Harry Liebersohn, Dr John Stuermer (First National Bank of Chicago), Professor Erich Wimmer (physicist, Universität Wien) and last but by no means least, Ms Renee Shiveley Leonard. They have corrected many pages of my writing, informed me on many topics, and encouraged me in my hope of producing works of interest and importance.

I am very grateful to Professor Michael Radnor, Professor David Roos and Ms Barbara Collins, respectively Director, Seminar Organizer, and Assistant Director of the Center for the Interdisciplinary Study of Science and Technology (CISST), Northwestern University, for their appreciation of my work. At the beginning, 'The Great Inertia', as a 30-page typed article, was the result of a talk given at CISST in May 1981. Again, in October 1983, I presented at CISST the last chapter of this book.

At the Twelfth Annual Meeting (May 1983, Buffalo) of the International Society for the Comparative Study of Civilizations (ISCSC), my work received strong encouragement. Professor Vytautas Kavolis (incumbent Chairman of the ISCSC) told me that *Comparative Civilizations Review* would consider publishing my introductory paper. After the meeting, Professor Burkart Holzner (University of Pittsburgh) and I discussed the topic many times, including exchanging several long letters. As a result of the discussion, I finished a summary, 'Science Development: Sino-Western Comparative Insights', which I started with a quotation from Ludwig van Beethoven, 'It had to go that way, if Herr Förster knew anything about composition', because I became more confident about the idea of a 'single pattern' of scientific development. I will never forget what Professor Cyrus Lee (Edinboro University of Pennsylvania) said at my presentation: '. . . You are telling us China's traditional difficulties. You are debating with acknowledged authorities. Yet in your speech itself, I think, we can see the hope for China.' Following the comment were another standing ovation and my tears.

In addition to the above-mentioned general encouragement from Professor Fairbank, I also want to acknowledge that I owe him a heavy intellectual debt. I have read his great work *The US & China* (4th edition) twice. I read it the first time four months after I had arrived in the US, when I still relied on a dictionary a good deal. Retrospectively I realised that the book must have influenced me in forming my idea of 'the great inertia'. What I remember vividly is that it reinforced my determination to pursue my 'proper study', namely studying China. Being fresh from Maoist, or neo-Maoist, China, and appointed by my university authority to pursue physics, I seemed to be experiencing a revelation when I read passages such as:[4]

> Among the educated Chinese upper class there was a strong and indeed inspiring tradition that the Confucian scholar had a moral responsibility to speak out against misgovernment. The famous maxim (Buddhist-tinged, to be sure) was handed down from a Song reformer: 'A scholar should be the first to become concerned with the world's troubles and the last to rejoice in its happiness.'

I reread *The US & China* a few years later when my understanding of China was much better than before,[5] and when the draft of this manuscript was well under way. This time, particularly in going through the first half of the volume, I was amazed at many illuminating comparisons and incisive analyses. 'That is wisdom!' I said to myself. 'And he really measures up to his avowed principle — anything worth saying can usually be said better more briefly.'[6]

In fact, almost no pertinent parallels, whether 'horizontal' or 'vertical', no vital causality, such as Chinese religions and authoritarianism, over-population as a root cause of many basic phenomena, etc. escaped his notice, although some parallels and connections he treats lightly, even facetiously. I need not list these comparative and causal analyses, but I must note one topic, which is the topic of my work. 'Why, Fairbank is writing "the great inertia"!' I exclaimed when I reread him. As soon as I had realised this I jotted down each page number and the subject by which I believe Professor Fairbank was explaining the Chinese inertia. I am also aware, however, that 'inertia' is such a provocative term that sometimes scholars argue for it in essence but avoid, or deny, it in words. Because I do not know Professor Fairbank's specific idea about using this term as a general and categorical characterisation for China, I prefer to append a note[7] to these remarks.

So far the persons who saved me from persecution in China and starvation — an understandable understatement — outside China are Mr W. Clement Stone, his relative Ms Barbara Stone, and his executive Dr Jack Early. I think the best way to express my deep gratitude to them is to quote Herbert J. Muller, an eminent historian who wrote prolifically on democracy and Western civilisation. Muller acknowledges in his *Freedom in the Western World* (1963): 'Since it is often said that our democratic society does not properly esteem its scholars, writers, and thinkers, I wish in particular to express my gratitude to the Foundations that help to support them better, perhaps, than almost any past society did . . .'

The CCP authority likes to make evaluations such as: in such and such a field the Chinese level is such and such years behind the advanced level of the world, and the time lags are generally sanguinely, or deceptively, put between 15 and 30 years. I would like to point out that in the most vital field China is more than 300 years behind the world's foremost, because people in advanced nations challenged and remonstrated with their political authorities. John Milton, for example, confronted the authority of his time and his land by declaring: 'Give me the liberty to know, to utter and to argue freely, according to conscience, above all liberties.' With *Areopagitica* Milton told his authority, who insisted upon licensing and censoring the press, that freedom of speech does not mean the toleration of evils, but means blessings that are essential to the life and happiness of a nation. Today, a Chinese can only shout these Miltonian statements in exile.

Telescopically speaking, the histories of all civil societies have traversed similar courses. Intolerance was common to all pre-modern states. Therefore, there was a need for a John Milton and his *Areopagitica* in order to change seventeenth-century England. That this did not happen in seventeenth-century China, eighteenth-century China, nineteenth-century China, and at least most of twentieth-century China, that is, that so far there could be no Chinese 'Milton' free from severe political persecution, is, in my judgement, another sure indication of the Chinese inertia.

My intrinsic optimism, which, let me pray, is more realistic than the optimism that prevailed in China more than six decades ago, however, reminds me that the world is on the whole changing towards being brighter and better. Currently there is a veritable, though disorganised — even saliently incoherent — great movement that diagnoses and cures China's chronic ailments. China is making progress; so is the Third World; and so is the whole world. If the eighteenth century was the Enlightenment of the West, the twentieth-century will hopefully be chronicled as 'the Enlightenment of the World'.

Since 'the Enlightenment of the World' is inversely proportional to 'the great inertia of the world', I would like to include here a short analysis of some major difficulties and trends. On the ideological level, we can identify a 'Western question' as well as an 'Eastern question'. Sometimes the demarcation is not clear cut (such as in some Islamic countries), but roughly speaking the 'Eastern question' applies to countries where one dominant ideological system rules with absolute political power, and no public and genuine dispute of different systems is allowed; the 'Western question' applies to the rest of world where dispute between different systems goes on in a whole spectrum of forms. (During the Second World War in some cases the 'Eastern question' occurred in the West and resulted in horrible bloodbaths.)

For many who believe that communist totalitarianism should be enlightened, or democratised, the 'Eastern question' means in practice the communist question, concerning which there exist some hypothetical questions of theoretical significance: Could a system of communist ideology free from political and academic arrogance have prevailed over the 'Eastern' peoples as Marxist communist ideology has done for these past decades? Could the leaders of the first victorious proletarian revolutions have started a political tradition other than one party (one coterie-one man) dictatorship in their socialist states? These are controversial questions. One possible theoretical solution to them is that both answers are negative; and historically, the communist interlude could not have posed a less severe 'Eastern question' than the one we have to face up to today. But, in actual development, communism has been changing. For three decades — since the death of Stalin, to be exact — economic revisionism and cultural flexibility have been a major remedial and driving force in all communist nations. Everywhere communists found that modelling themselves on capitalist management, respecting individual incentives, assuming entrepreneurial responsibility, etc. are a necessity, and sticking to socialist egalitarianism leads to disastrous results. Ideologues have become sceptical of their classical doctrines.

The 'Western question' is a traditional one; in the eighteenth century, during the Enlightenment, Voltaire and others squarely confronted this question. Today I would like to phrase it as 'the rationalisation of religious thoughts'. Galileo's recent exoneration, the equal-time allotment to evolutionists and creationists in some American states, the debate over public prayer, etc. are ripples and vortices in a ceaseless stream. In the Middle East, Northern Ireland, and some other areas, terrorism — the combination of extreme violence and irrationality — has become the political and military manifestation of the 'Western question'. And the demand of its resolution in these clashing areas is certainly no less urgent than any other questions.

There are not many years left before we use up this century. Are we truly entitled to be optimistic?

Ideological disputations will and should last, yet the changing contents and manner of the disputation are the major indicators of civilisational progress.

The harshness of the word 'defection' (I may still retain some traditional Chinese ethical values), and the extreme personal difficulties involved often pushed me to soul-searching self-analysis. So far my conclusion is that the following four facts form the basis of my political decision and intellectual efforts:

(1) The training in natural science inclined me to objectivity and to using independent reasoning.

(2) Maoist revolutionary and nationalist propaganda, on its surface, resonated with my patriotic passion.

(3) As a corollary, this propaganda also roused in me bitter hatred against its hypocrisy when I saw through communist totalitarianism.

(4) The sharp Sino-Western contrast in material and intellectual achievements inspired as well as humiliated me.

In the past few years sometimes I also had difficulty in conversing with 'progressive' Western people. Gradually I learned how should I talk to them:

(a) If you resent inequalities in monetary income, career opportunity, and social influence, then you certainly hate oppression. Therefore, if you were in my Chinese shoes, you might very well share my feelings.

(b) You remind me that there may be great merit in traditional Chinese values and institutions, that it may be wise to keep them, and that the CCP may be just doing that. Now I have told you the causes (1), (2), (3) and (4) that force me into my present way of thinking, but these may very well be the common experiences of the Chinese people, particularly of the younger and coming generations. Especially point (1) and point (4). The CCP wants to industrialise and 'modernise' the nation. Tens of thousands of engineers, doctors, and scientists are being trained every year. But all totalitarian rulers are faced with an ineradicable headache: How can a double standard be maintained among people so that they think objectively and independently in their professional fields but also obey blindly and slavishly on many vital problems concerning their public and private life? Today's world is shrinking rapidly with the progress of telecommunications and mass transportation. How can a selfish and narrow-minded regime control its people so that they do not admire, envy, and emulate a material life which is on the average 15-25 times higher, or a much more profound, eclectic, varied and creative intellectual tradition? This is another ineradicable headache for those powerful hypocrites sitting in the Prohibited Palacc in Beijing.

Once when Confucius and Zi-lu were travelling, they saw two scholar-recluses ploughing. Confucius told Zi-lu to ask them the way to the ford.

When one of the recluses learned that Zi-lu's master was Confucius, he simply refused to communicate with them, saying, 'He should know where the ford is then.'

When the other recluse learned that Zi-lu was a disciple of Confucius, who must have been known for taking an interest in current affairs, he admonished Zi-lu to follow a recluse's outlook: 'The whole world goes its way and who is to change it? Better, surely, to follow those who shun the world than the one who shuns certain men!'

Zi-lu returned and related the conversation to Confucius, who commented ruefully: 'Birds and beasts are no company for men. If the world were on the right path I should not try to change it.'[8]

Like Zi-lu, I was admonished to change my course.

Also like Zi-lu, I decide to follow Confucius, who thought the world was not on its right path, and who wanted to contribute his effort to its changing.

Notes

1. I finished drafting this 'Explanations and Acknowledgements' in August 1983, immediately after rewriting and organising the four articles. Now that the collection is going to be published as a small book, I am much honoured that two distinguished scholars promised to write forewords for me. It is very understandable that Professor Fairbank's appreciation (and/or criticisms — since I still have not seen his Preface) of the work especially thrilled me. Yet when I mailed my MS to the Fairbank Center of Harvard in August 1983 I did not include the 'Dedication' and 'Explanations and Acknowledgements', because I feared that the strong political

overtones in these two pieces of writing might mar my image as a serious scholar (Qian, March 1984).

2. Carl L. Becker, *The Heavenly City of the Eighteenth-Century Philosophers* (Yale University Press, New Haven, 1932; 1975, 39th printing), p. 141.

3. According to what I know, this principle was Lev Tolstoy's invention; as a guide to historical scholarship, again according to what I know, it is G. Parker's discovery: *The Army of Flanders and the Spanish Road* (Cambridge University Press, Cambridge, 1972), p. ix.

4. J. K. Fairbank, *The US & China* (4th edn) (Harvard University Press, Cambridge, Mass., 1979), p. 134.

5. This sounds ironic, yet the reason is extremely simple: no Chinese authorities have ever allowed people to study China objectively and impartially. In other words, in China history has never been an arguable and debatable scholarship — please don't be fooled by Mao's 'Let a hundred flowers bloom together.'

6. Fairbank, *The US & China*, p. xvii.

7. Examples include:

'. . . a vicious interdependence between dense population and intensive use of soil . . . Once established, this economy developed its inertia . . .' (p. 13).

'Of course massed manpower built the Egyptian pyramids and other wonders of ancient empires, but only in China has this custom persisted to the present day.' (p. 17)

'The chief social division has therefore been that between town and countryside . . . This bifurcation still underlies the Chinese scene and makes it difficult to spread the control of the state from the few to the many.' (p. 20)

'China has been the stronghold of the family system and has derived both strength and inertia from it.' (p. 21)

'In his [Wittfogel's] analysis they all shared common features of universal kingship, land-based economy, generally ideographic writing, and a monolithic bureaucratic government that dominated large-scale economic activity so that no sanction for private enterprise became established. But of them all, only the Chinese empire survived into modern times.' (p. 28)

'Many of these "oriental" characteristics of the early Chinese empire lasted down to recent times.' (p. 30)

'Until recently the form of these characters had hardly changed since the early Christian era. Until a generation ago the Chinese were still getting along with a written language comparable in age to ancient Greek and Latin . . .' (p. 39)

'. . . a composite ruling class . . . The structure was flexible, automatically self-perpetuating, and very stable.' (p. 46)

'. . . its esteem for age over youth, for the past over present, for established authority over innovation, has in fact provided one of the great historic answers to the problem of social stability.' (p. 53)

'For most of two thousand years the Confucian ideology was made the chief subject of study in the world's largest state. Nowhere else have the sanctions of government power been based for so many centuries upon a single consistent pattern of ideas . . .' (pp. 53–4)

'For more than 2000 years this system of territorial bureaucracy has been epitomized in the walled administrative city of the *xian* or county.' (p. 61)

'For more than a thousand years, down to 1905, this imperial institution [the examination system] produced administrators who had established their qualifications for official life by thoroughly indoctrinating themselves in the official orthodoxy . . .' (p. 62)

'The history of technology in China confronts us with still another paradox. Until late medieval times China was at least on a par with and in many ways ahead of Western Europe in technological inventiveness and systematic observations of nature . . .' (p. 74)

'. . . Chinese institutions . . . over the centuries had developed great self-sufficiency, balance, and stability . . . The deep-inlaid inertia of the Confucian pattern in Chinese political life during two millennia explains why China's modern revolution against it has been so grievously long-drawn-out.' (p. 79)

'In this way the very ancient idea of universal kingship, used by early rulers in Mesopotamia, Egypt, India, and even pre-Columbian America, was able to persist in China down to modern times.' (p. 100)'

'This nondevelopment of Chinese law along lines familiar to the West was plainly related to the nondevelopment of capitalism and an independent business class in the old China.' (p. 123)

'. . . the inertia of the traditional Chinese polity — it could not be Westernised but could function only on its own terms, which were growing out of date.' (p. 191)

'. . . China's complacent self-sufficiency and inertia . . .' (p. 203)

'This long established low-level manpower economy was perpetuated by the Chinese emphasis upon the virtue of contentment and limitation of wants.' (p. 271)

'Indeed the narrow horizon, low efficiency, poor diet, and chronic diseases of the Chinese peasant, which struck the eye of the modern investigator, were always an integral part of the old society, just as they were part of premodern society in Europe.' (pp. 271–4)

'Chinese governments over the centuries, struggling to maintain a unified state and a uniform culture over a vast and diverse subcontinent, became skilled at imposing central controls and large scale bureaucratic administration . . .' (p. 359)

'Muscle power had always been the country's chief natural resource . . . the grinding down and slow extinction of life through prolonged and ill-equipped labor, always a part of the Chinese scene . . .' (p. 379)

'. . . the immovable Middle Kingdom . . . authoritarian tradition could be invoked for modern purposes. Thus Confucianism in one of its aspects has a certain resonance with Marxism.' (p. 383)

'Communist self-criticism was reminiscent of the Confucian doctrine of self-cultivation . . .' (p. 384)

'This unitary system has worked in the People's Republic as under imperial Confucianism because moral-ideological authority and political power have been combined at the top . . .' (p. 392)

'[The policy of let a hundred flowers bloom together, let the hundred schools of thought contend] would be in the imperial censor's tradition of "loyal remonstrance," not like a "loyal opposite" in the West.' (p. 402)

'A candidate might go through this application procedure two or three times before winning acceptance, not unlike the repeated efforts of scholars to pass the examinations in the old days.' (p. 429)

'Watching skilled workers make molds in a Honan iron foundry, one remembers that cast iron was in fact invented in the region in ancient times, many centuries before it was known in Europe.' (p. 448)

[Finally there are instances in the section "China Today in the Light of Her Past" (pp. 462–70), which tells us that 'continuity and discontinuity go together as the warp and weft of history'.]

8. Adapted from Sima Qian (145-*c*.89 BC), 'Confucius' in *Shi Ji*.

1 THE GREAT INERTIA: AN INTRODUCTION[1]

Ancient and continuous, accomplished and autonomous, the Chinese (or Han) civilisation has been recognised by many experts as tardy in development. In the past several decades, historians have demonstrated that science was an inseparable part of the Chinese civilisation, and some of them thought that it shared the common fate of tardiness. This work argues that scientific development cannot be historically isolated from civilisational development as a whole. To put it otherwise, the growth of modern science was an organic process in a society's holistic undertaking. There are, however, sharp differences of opinion on the history of science and technology in traditional China. Joseph Needham thinks that it is completely wrong to use words such as inertia or stagnation to describe China's scientific and technological past, and that physics might have developed totally differently. Nathan Sivin asserts that there was already a 'scientific revolution' in seventeenth-century China, and that with regard to developing modern exact science China was comparable to France. An overall evaluation of the status of science in traditional China must imply camparisons between China and Western Europe, both in socio-political evolution and in the development of physical sciences. The debates around the use of the word 'inertia', the alleged Chinese 'scientific revolution', a hypothetical different history of physics, the pseudo-physical basis of the 'organicist' philosophy, etc. lead to some very interesting and significant theoretical (certainly still controversial) insights, namely:

(1) a criterion for the historic development of science, based on the observation that, compared with other fields of culture and institutions, science possesses a 'uniqueness' and 'uniqueness in historic development', rooted in the scientific requirement of 'immediate' empirical verifiability or rigorous logic (a principle that sorts out different fields within science itself according to their degree of 'uniqueness': there were only a few contending physical theories in each branch of physical phenomena before 'classical physics' established a unifying theory in a 'classic' manner; but in life and geological sciences, contending theories coexisted very

much longer);

(2) the 'theorem' that 'software' decides (in comparison with the existing technological and scientific conditions, ruling ideology and political institutions played a dominantly important role in a traditional society's process of modernity);

(3) the smothering effect of a unified system that dominated a vast (and densely populated) territory;

(4) the 'principle of the degree of intellectual activation' that should guide us generally in studying modernisation problems;

(5) about the guiding scientific methodology that is incorrectly hypothesised by some historians as being 'rectified'; and some other propositions.

This work is meant to be a causal inquiry into the Chinese stagnation. It is my high hope that these theoretical insights, which I have arrived at during the course of inquiry, will find some more applications in tackling Chinese problems of a more universal character.

Recently in China there appeared in one of the leading philosophical periodicals, the *Journal of Dialectics of Nature* (Spring 1983), a special section discussing 'the causes of China's scientific and technological backwardness in the recent past'.[2] I take no personal credit for this event, but I am very pleased about it, since after more than 30 years of blind self-glorification, this small official reorientation broaches a genuinely significant, indeed uniquely important question for China. If we trace the matter further back we may say that this is not a small reorientation, but rather a departure from the long-established tradition in all Communist nations.

In this collection, Essay One and Essay Two (Chapters 2 and 3) discuss respectively the factual and the logical aspects of Chinese scientific inertia. In fact, I finished Essay Two first and then I turned to a detailed Sino-Japanese comparative study. Essay One came out because my advisors urged me to go through all Needham's fat volumes and finish some serious study.[3] Reluctantly and fortunately I followed the guidance of senior historians. As a result, this collection emerged unexpectedly as a more or less self-sufficient work.

Essay Three (Chapter 4) is a shorter one, in which I want to put philosophical arguments on some basic and solidly-established physical facts. Needham thinks that traditional Chinese science was, and will be, marvellous. No one denies that traditional Chinese science did not develop, or contribute directly to, modern science, which is evidently characterised by mechanistic clarity and accuracy. But Needham stressed that ancient Chinese philosophic thought, the so-called 'correlativism', will usher into a certain futuristic scientific methodology, the organicist philosophy.[4] However, in the second volume of his famous series, Needham consistently uses a false understanding of a field force, the relativistic idea of the order of consecutive events, and some basic mathematical concepts to support his philosophical faith. I think his misunderstandings are not isolated errors, and it may be worth while to put down my critique in black and white. It is not that I am eager to grasp any opportunity to denigrate either Needham or traditional Chinese scientific thinking; rather, it is because I am a 'theoretical' reductionist[5] and I do not share Needham's faith in the organicist philosophy.

Whenever I read in Needham's works a Chinese passage which talks about the Yin and Yang, or what Needham names as the Pattern, the Order, or resonance that embody the Tao, I knew that he would praise it as very scientific, very illuminating, while I would frown at the same passage. But I have not lost my sense of history and taken an anachronistic, modern point of view. I frowned because of history, not in spite of it. The so-called Needham Puzzle is this: with so much marvellous achievement, why did not medieval China develop modern science? My work started by pursuing this Puzzle, and my conclusion is that different lines of argument lead to the resolution (or rather the invalidation) of the Puzzle: the pattern of China's past; the difficulty of China's present; and the status of traditional Chinese science itself. When I looked at the phenomenology of Chinese science, as voluminously presented in Needham's works, I frowned. How could this essentially pre-modern body of thought become modern without the necessary examination and re-examination, criticism and counter-criticism, formulation and reformulation, mathematisation and re-mathematisation over many generations? Did traditional China at any time offer the necessary politico-ideological conditions to sustain the corresponding degree of intellectual activation and creativity? In other words, I doubt that China has ever provided a social environment that would engender the necessary competition of creative thinking and objective criticism, or the constant re-examination through empirical testing and rigorous reasoning.

I frowned because of history. The actual (and only) Scientific Revolution was a triumph of mechanistic thought. By analysing the inherent relations between various fields of physics, it becomes clear that physics could not be modernised otherwise. Consequently, in terms of scientific contents, the Scientific Revolution — at least the 'mechanical revolution' and the 'physical revolution' — could not have emerged in a different order. In this way we can establish an objective criterion for scientific 'modernity',[6] which is neither chronological[7] nor based on verifiability. By this criterion, not only Euclid and Archimedes were the forerunners of modern physics, but even the mistaken Aristotelian 'law of falling bodies' was 'modern', because it was clearly formulated and subject to quantitative testing. Again, the archaic Ptolemaic planetary astronomy was 'modern', because it correctly isolated the solar system and attempted orbital explanations. Also by this criterion, the apogee of traditional Chinese science was not magnetism, as Needham assumes, but the theoretical study of the musical scale, which was crowned by Zhu Zai-yu's invention of the equal temperament in 1584.[8]

In fact I think one of Needham's major speculations is untenable. He states that since the Chinese were so much in advance of the West in the study of magnetism, had the social conditions been favourable for the development of modern science in China, then the Chinese might have pushed into field physics, without going through the stage of 'billiard-ball' physics.[9] This argument fails to convince because the concept of a field of force in modern physics is based on none other than the Newtonian concept of 'billiard-ball' interactions: fields are physical objects; fields interact with particles or other fields in accordance with mechanical laws, either Newtonian or quantum. The statistical nature of quantum physics by no means negates the 'law of physical causality'.[10] So long as we pay casual attention to various types of photographic records of elementary particle interactions, we would admit with alacrity that down to the depths of the material world, very much deeper than the ultimate mechanisms that decide the essence of life, billiard-ball interactions prevail, indeed exclusively.

A Theoretical Discourse

More than 60 years ago, Fen You-lan, then a young man working at Columbia University, finished an inquisitive and provocative study on a macro-historical topic. His essay, 'Why China Has No Science — an Interpretation of the History and Consequences of Chinese Philosophy', was published in 1922 in the *International Journal of Ethics*. It was only natural that this work was done after the so-called Chinese Renaissance, in the wake of the First World War, and by a Chinese student who was then staying in the West. In the same year, 1922,[11] Bertrand Russell published his book *The Problem of China*, which he started by asking: 'Can Chinese virtues be preserved?'[12] And he also told us: 'The distinctive merit of our civilization, I should say, is the scientific method; the distinctive merit of the Chinese is a just conception of the ends of life.'[13] Meanwhile, Feng You-lan concluded his own work by declaring: 'Anyway, the Chinese conception of life may be mistaken, but the Chinese experience cannot be a failure.' Eastern iconoclasts and Western iconoclasts sometimes cannot see eye to eye on what is to be valued,[14] and this problem is still with us.

In retrospect, what appears to be more ironic is that after these 60 years — during which dozens of nations changed beyond recognition — Feng You-lan's basic question, 'What keeps China back?' is now raised again, while China's percentage of rural population is still among the world's highest, while its level of *per capita* income and political enlightenment are still among the world's lowest. This can be explained briefly. The 60 year span divides into two parts: in the first half China was busy fighting, either civil wars or the Second World War; in the second half China was busy boasting (about Mao's 'achievements'), as well as fighting (following Mao's instructions), and any study of China's traditional inertia would comprise a serious political offence. This period, though short, was in fact a characteristic extension of the two-thousand-year traditional China (from the beginning of the second century BC to the middle of the last century, when autocratic unification was basically a reality and continuously an ideological ideal).

Feng You-lan's basic assumption was that 'What keeps China back is that she has no science.'[15] His paper concentrates on answering the question why China had no science. Apparently Feng's questioning sounds similar to the Needham Puzzle that I mentioned before — why did not modern science emerge in China? But these two schools arise from opposite points of view. Feng's, to which I belong, has embarked on a causal inquiry into China's long stagnation, whereas the Needham school pours its energy into summarising everything that smacks of science in Chinese history,[16] partly in order to correct a 'total misconception' about China. The former raises questions out of frustration; the latter, out of admiration.

Needham states that it is absolutely wrong to use words like inert, stagnant, or changeless to describe China. Instead, he favours another set of adjectives: homeostatic, cybernetic, in other words, self-regulative. He tells us that Chinese society had a continuous tendency to return to its original character after each disturbance. Chinese historians have long taken notice of the political aspect of this phenomenon, the dynastic cycles. Traditional China was indeed characterised by a pair of macro-historical[17] opposites: internal and political instability versus institutional and ideological continuity. It is precisely this fundamental continuity that justifies the application of the term inertia to traditional China. Many scholars, Western as well as Chinese, particularly those who are anxious to drive China out of its inert state, take a comparative or world-wide outlook, and describe China with such critical terms as inertia and stagnation. Ironically, although Needham is doubtless one of the scholars most familiar with the characteristic Chinese inertia,[18] he asserts it is absolutely wrong to say so.[19]

On a question like this, or the even more sensitive yet less definable question of national character, both foreigners and nationals are caught between sensitivity to national pride on one side and a temptation to say 'the emperor is naked' on the other. The balance between these two sides, which differs from person to person, contributes to a scholar's selection, evaluation, and presentation of historical facts and interpretations. None the less, I choose to work this way, because I believe that in the long run, there is no contradiction at all between being self-critical and being patriotic. Therefore, it is understandable that I really have some reasons to be pleased when I read in Chinese papers the contents of the latest issue of the Chinese philosophical periodical the *Journal of Dialectics of Nature*.

On the question of basic Chinese retardation I again agree with Feng You-lan in looking for 'software' instead of 'hardware' causes.[20] By 'hardware' I mean technological hardware and hard sciences. In Feng's work — although his generation did not use these terms — 'software' was China's guiding philosophy and social values, but my answer to the question of non-development differs from his, in both basic viewpoint and logic. Feng offered a purely philosophical solution to China's non-development of science. He thought the long delay was rooted in the introspectiveness of the dominant Chinese philosophies. To him it was because in traditional China intellectual curiosity was not directed to the material world, but was engaged in finding good and happiness in the human mind.

I do not agree with Feng's basic assumption, that 'What keeps China back is that she has no science', even if 'science' is now appropriately replaced by 'modern science'. Not only is this a simplistic formulation, but it has also put the cart before the horse. In comparing China's 'backwardness', or the Chinese tardiness, with 'the lack of modern science', I would not take the latter as the cause and the former the effect. If Feng were to hear this statement he would most likely say that he does not identify it as the true cause, and I agree. But I understand the non-development of modern science as only one of the symptoms of the Chinese ailment; the other major symptoms are the non-development of modern economy, modern politics, and modern culture. Secondly, I do not identify the root cause in the character of China's dominant philosophies.

One salient feature of the dominant Chinese philosophies, Confucianism, Taoism, and Buddhism, is that they were so near to each other — or they were made to fall into a narrow and self-consistent scope — whereas the schools of Mohism, the Logicians, and the Legalists, which had flourished in the pre-Qin-Han era, were all left in oblivion for milleniums. Another salient feature of the dominant Chinese philosophy, usually simply called Confucianism, is that it is exclusively politico-ethical. Logically these two aspects were linked: a predominantly political ideology is an intolerant ideology. In practice, for the dual purpose of social and intellectual control, it was invented, consolidated, and perfected over centuries the great Chinese bureaucracy. The central mechanism of Chinese bureaucratism was its system of training and selecting officialdom — the time-honoured Chinese system of civil service examinations. The beginning of this system traces back at least to the Sui Dynasty (late in the sixth century); it ended in 1905, the same year in which Western Europe saw the publication of Einstein's special relativity, a work of the same historic status as that of the *Principia*. In other words, compared with Western Europe (in Section 2 of Essay One we will justify the comparison),

'Confucian' China had missed at least two 'Scientific Revolutions'. Since 1313, the required reading for every aspiring learned man was officially limited to a set of Neo-Confucian texts; the contents of the written examinations — unlike the situation in the Tang and Song Dynasties — became incredibly narrow, and the style of the standard prose ridiculously stale.

In terms of vastness and population China was comparable to Western Europe as a whole. Some scholars emphasise the geographic differences between these two sides. Yet I think the decisive difference was institutional. Unlike Western Europe, traditional China was imperially unified. Although few dynasties were truly powerful and stable, the continuation of a unified ideological control was guaranteed by the persistent ethical-ideological values and political institutions. Historians may identify the Tang and Song Dynasties as a deviationist period of Medieval Flowering (the eighth to thirteenth centuries), but still traditional China remained a basically uniform and intolerant politico-ideological environment. Its political intolerance prohibited any change in institutions themselves; its economic intolerance barred the growth of capitalism; its intellectual intolerance hindered the growth of a more pluralistic culture and science.

Mainly owing to the institutional difference, the two ends of Eurasia traversed very different patterns of historic development. In Europe, in sharp contrast to Islamic and Chinese traditions, the attempt to combine politico-military power and ethical-ideological orthodoxy never achieved much. Relatively speaking, the Dark Ages in Western history had been short. After that one movement after another challenged the traditional European order, a combination of feudal militarism and ecclesiastical obscurantism. One basic reason why these movements have been remembered as epoch-making events was that they were effective in reducing European intolerance and in diversifying Europe's intellectual nourishments (with ancient and foreign wisdom, and gradually, with creativity). Although the tension between intolerance and enlightenment was generally donned with a religious cloak, the potentiality that was latent in the tension's step-by-step resolution was extremely rich in content. Since the sixteenth century, modern science began to emerge in a Western Europe that was changing in every aspect of its life.

In short, my answer to China's non-development question is

essentially political, or politico-ideological. The major lesson that I derive from these Sino-European comparative observations is that the Chinese politico-ideological rigidity — its uniformity over a vast territory fed back strongly to intensify itself — effectively prevented corresponding progressive campaigns to occur in China. The Christian view towards the individual, the Christian attitude towards nature, and sometimes the Christian interest in the Greek legacy, were all important to Western development, and to scientific development in particular. Conditions such as these should also be grouped under my Sino-European politico-ideological contrasts.

This theory must look conspicuously unorthodox by present-day Chinese standards. Recently I read a 700-page Chinese book, *The Scientific and Technological Achievements in China's Past*, compiled by the Institute of the History of Natural Sciences of Academic Sinica, published in Peking in 1980. Its Preface points out that:[21]

> Except for short periods, China has for thousands of years kept its political unification. Although a multi-racial nation, China has never been divided, such as Europe has. We are even less like the Roman Empire or the Mongol Empire which decayed after a short-lived power . . . The Chinese nation has been in existence for 4,000 years on the earth and continuously growing stronger and more prosperous. One of the major reasons for this strength, as is embodied by the content of this book, is that we have our glorious scientific and cultural achievements.

Needless to say, this book still belongs to the category of self-glorification. The interesting thing to me is the reverse logic in this paragraph. I think the last phrase, 'glorious scientific and cultural achievements', should be translated into 'powerful ideological and effective bureaucratic control', that maintained China's institutional continuity, ideological rigidity, and cultural monotony. It was China's imperial and ideological unification that arrested China's scientific and cultural development. China's long unification demonstrates not the strength of China, but the strength of the power that ruled China. In sum, the two contrasting arguments are: one says China's glorious traditional science and culture guaranteed the long existence of the Chinese Empire, whereas the other says that the factors that sustained (not very successfully) the

Chinese Empire blocked the appearance of modern science and a richer culture in China.

As I have mentioned earlier, the Chinese dynastic cycles show at once political instability and ideological continuity. Hundreds of peasant rebellions, civil wars, nomadic invasions, and dynastic shifts belie a true unification. Except for a few mighty dynastic periods, China was not truly powerful, because its people were on the whole neither economically nor intellectually powerful. European people were comparatively powerful; and that is why no single power, military, political, or spiritual, has ever succeeded in ruling them all. In so far as we are talking about the emergence of modern science, what is wrong with a disunited Europe? From Copernicus to Newton, in a long list of great creative scientists, no two standing side by side had, as a rule, the same nationality, or the same religious denomination.[22] The situation shows the extensive and intensive degree of intellectual activation in sixteenth- and seventeenth-century Western Europe.

The comparison between China and Western Europe revealed to me other analogues: the comparison of traditional China with ancient feudal China, the comparison of traditional China and ancient Greece, and the comparison between late traditional China and centralised-feudal Japan. These parallel contrasts suggest that the competition among a cluster of feudally or otherwise independent states, which share a common culture, was a great stimulus to development. The Warring States period of pre-221 BC China was recognisably a dynamic era in many ways, as it is stated clearly by John King Fairbank, Carrington Goodrich, and others. During that time learned men and craftsmen lived in a much more stimulating atmosphere than later, in imperially unified China. Science and culture could not guarantee China's unification. Rather it was a dynastic China, ideologically as well as politically unified, that underpinned the long-lasting Chinese traditionalism and archaism. Had the condition of the 'One Hundred Contending Schools' lasted several centuries longer, positive scientific results might have reached a much greater accumulation and the leading Chinese thought might have risen to a much higher level.

We still do not know to what extent Greek intellectual excellence depended on characteristic Greek competitiveness. We know about the political and military hostilities among the ancient Greek city-states, and their consequent development. We also know about the tradition of the Olympic Games. We are therefore tempted to assume that a kind of uninstitutionalised perennial 'intellectual Olympic games' was responsible for the eminent heritage. Besides, competitiveness should be understood generally. When Aristotle left Athens and established his own school, he was looking for better competitive conditions for the sake of his scholarship. If one is mystified by the prodigious accomplishments of the logical structure of Euclidean geometry, then only by reading some Greek philosophical and ethical works prior to Euclid can one reduce the feeling of mystery and awe. Besides, a learned man competes with himself, once the 'chain reaction' of intellectual creativity is ignited.

The Chinese people usually describe the Sino-Japanese proximity as 'yi yi dai shui' — only one narrow strip of water (that separates the two nations), and 'tong wen ton zhong' — the same race and the same writing system. It requires true insight to overlook superficial similarities and reach a recognition of their essential differences. As Edwin Reischauer points out:[23]

> By the 12th century Japan was on the threshold of an even greater departure from East Asian norms. This was the development of a feudal system, which over the next seven centuries was to go through phases that had many striking parallels to the feudal experience of Western Europe between the 9th and 15th centuries.

Hence none of the numerous similarities between China and Japan such as racial identity, geographical proximity, Confucianism, Buddhism, rice agriculture, the self-sufficient agrarian economy, the writing system, the many mutual cultural borrowings, etc. were decisive in the two nations' respective social metamorphosis; but in the different politico-social structures of the two societies lay the potentials for their respective future development, potentials which were realised in the nineteenth century when both countries faced the threat of Western imperialism.

If we compare the contemporaneous Meiji Restoration and the Chinese Westernisation Movement in the 1860s, we have a good illustration for our theory that 'software decides'. In Japan, modernisation was carried out both in political institutions and economic constructions, while in China the Manchu regime looked to the West for only superficial technological imitations with the sole purpose of maintaining its outmoded political establishment.

Small wonder the results on the two sides emerged in sharp contrast.

The relevance of these general comparisons to the rise of modern science is obvious, because I believe, like Needham, that modern science is an inseparable part of general modernity, which also has its economic, political, cultural and demographic aspects. Given adequate social conditions, the growth of science itself depends on the impetus for systematisation and explanation of phenomena, on the search for causes behind effects, and on the reasoned organisation of explanations conceived according to one's perception of rationality.

A plain fact is that the progress of science depends on people's doing science, which in turn depends on political tolerance and social encouragement. According to this understanding we may raise a series of questions in order to examine traditional China's potential to develop modern rational science, or to compare China with Western Europe in their efforts put — not necessarily consciously — in this direction. Finally, this line of thinking led me to the idea of a general criterion, which I call the principle of the degree of intellectual activation and creativity.[24] Science is such a great and difficult enterprise, that every creative scientist must be working in an interactive milieu, including interactions with accumulations from the past. There is no question that a reasonably large net of communication is more advantageous than a limited one. On the other hand, since in the past science had been the spontaneous engagement of only a few *philosophers*, the size of the community is in fact not as important as 'the degree of intellectual activation': How many philosophers were there in every 10,000 people? How many different systems of thinking were available to them, and how wide was the range of tolerance within which they were allowed to venture their imagination? How much probability was there that their intellectual creations would be carried on by upcoming philosophers? These questions comprise the extensive, intensive, pluralistic and inheritable aspects of intellectual activation and creativity.

Modernity in general is a great and difficult enterprise which requires tremendous efforts from the whole society. In earlier times, from the Renaissance to the Industrial Revolution, such efforts might not be organised for or conscious of their historic mission, but, in their overall contribution to the growth of modernity, they are constructive, innovative, explanatory (intellectual), or exploratory (experimental).

This principle, or criterion, is phenomenological, because it admits that more basic factors work behind intellectual activation and creativity on a societal scale: in certain specific ways some factors promote creativity, while in certain specific ways others inhibit it.

The Sino-European contrast in political structures helps to illustrate this principle. First there is its extensive aspect. Although China had the tradition of 'promotion through examinations', there were not many universities in China — if 'Guo Zi Jian' can be called the imperial university. In Europe, kingdoms, city-states, bishoprics, and principalities took on responsibilities for a number of institutions. As soon as universities started in Western Europe they spread rapidly. Ideological tolerance and cultural receptivity are postive factors for another aspect of intellectual creativity: pluralism. Since in the six centuries preceding 1905 the Chinese examination system required nothing but a small collection of Neo-Confucian classics, the detrimental influence on mobilising, or activating, intellectual creativity was fatal.

In the last few decades of the sixteenth-century, Hideyoshi, one of the founders of the Japanese Tokugawa regime, became so ambitious that he waged full-fledged invasions against Korea and China. A present-day reader may puzzle over the material foundation of his seemingly unrealistic ambition. By some more reading, one found that since the arrival of three Portuguese sailors in 1543, Japan had made rapid progress in manufacturing light firearms. Firstly, Tokitaka, the daimyo who first saw the Portuguese sailors, spent a large sum of money to buy a pair of arquebuses from them. Then he immediately asked his swordsmith to copy and produce the imported weapons. Other daimyo also showed great interest in arquebuses and other types of firearms. Amazingly enough, within only twelve years after 1543 Japan had become a manufacturer of firearms and was exporting them to southeastern Asia!

China's slow response to light firearms offers a revealing contrast. Using the principle of the 'degree of intellectual activation', we can put the contrast in parallel with the difference between a provincial governor in imperial China and a daimyo in feudal Japan — the former a hired manager, the latter the lord of his own territory. In China a scholar-official ascended along the classic status ladder, xiu-cai—ju-ren—jin-shi, by competing with his peers in memorising classics and composing 'eight-legged-proses'. On the other hand, the life-and-death strife between Japanese feudal lords demanded that they be prepared to adopt and use any practical means. In short, the fundamental Sino-Japanese differences are great. As a result, the two countries have been different in the degree of intellectual activation and creativity achieved.

Once formulated, this principle appears both self-evident and useful. It serves us as a clear-cut criterion. In examining problems that relate to the growth of modern science and modernity in general, it helps us to identify and analyse advantageous and disadvantageous factors and conditions.

Needham has written a perceptive passage describing China's scientific stagnation:[25]

> The only trouble about the Chinese five-element theories was that they went on too long. What was quite advanced for the +1st century was tolerable in the +11th century, and did not become scandalous until the +18th. The question returns once again to the fact that Europe had a Renaissance, a Reformation, and great concomitant economic changes, while China did not.

During the millennium after the ninth century, the evolutionary pattern of the history of Western Europe contrasts notably with that of traditional China. At the western end of Eurasia, there were the Carolingian Renaissance, the spread of universities, the incorporation of Aristotelianism into Christian theology, the translation movement, the Renaissance, the Reformations (both the Protestant and the Catholic), the Scientific Revolution, the Enlightenment, and the Industrial Revolution. At the eastern end of Eurasia, there was an equal number of, if not more, dynastic shifts, rebellions and invasions. Hu Shi, however, in his 1933 Haskel Lectures in Comparative Religion, said that since the Tang Dynasty there had already been five discernible 'renaissances' in Chinese history.[26] (Now I believe scholars on both sides of the Taiwan Straits would like to add a couple more to his list.) But Hu Shi also analysed the common defects of those pre-1919 Chinese renaissances. Judging by the principle of degree of intellectual activation and creativity, I would say that he had only paid attention to the inheritability of those movements — or more exactly their lack of inheritability.[27] What traditional China unfortunately lacked were a series of totally new intellectual elements: new attitudes, new ways of thinking, new fields of interests, new epistemological standards,[28] and most importantly a propitious politico-social condition to produce, sustain, and promote them. From time to time there appeared more or less isolated geniuses, yet they were not enough to sustain a genuine continuous Renaissance. Hence, in the light of the principle of intellectual activation and creativity, we see that the *developmental* Sino-European contrast illustrates, even better than its *structural* counterpart, the characteristic Chinese inertia.

Notes

1. Based on a presentation at the 12th Annual Meeting of the International Society for the Comparative Studies of Civilizations (ISCSC) at Buffalo NY, 26–8 May 1983. The presentation was thereafter published in the *Comparative Civilizations Review*.

2. *Zi-ren Bian-zhen-fa Tong-xing (Journal of Dialectics of Nature)*, vol 5, no. 2 (Beijing, April 1983), pp. 38–57.

3. In fact I have not yet had time to finish a serious study of all Needham's published volumes. The original topic of my Essay One (Chapter 2) that I had planned to write was 'The Phenomenology of Scientific and Technological Stagnation in Traditional China'. Since I could not finish working on Needham's and his collaborators' works on technology and engineering, the scope of Essay One is appropriately limited to physical sciences. This schematic change appears advantageous in hindsight, since at the moment, in the scholarship of the history of science, the difference between science and technology, not their connections and unity, should be stressed.

4. See Essay Three (Chapter 4), particularly the quoted passages identified by notes 7 and 8.

5. In a talk with Professor David Joravsky about Essay Three, he suggested that I read R. C. Lewontin's 'The Corpse in the Elevator', a recent article in *The New York Review of Books* (20 Jan. 1983), to acquaint myself with some recent challenges to the mechanistic philosophy in life sciences. I did, and found that, as Lewontin's reader, I can actually understand '*the corpse* in the elevator' in just the opposite way from its author.

Here, in this article an eminent, non-reductionist biologist tells us that he believes:

> Whether reductionist, holist, or dialectician, all parties agree on a materialist view that everything in the world is some manifestation of matter and energy . . .
>
> Yet every materialist would have to admit that if I could build with transistors or wires or whatever hardware an exact copy of my brain, and then set the electronic currents flowing in it, as they are now flowing in my brain, that machine could have written this sentence. That is, it could have contemplated its own existence.

So, a non-reductionist is reductionist enough. From these statements, life scientists like Lewontin would agree that an effect is, in the final analysis, caused by a physical agent. In other words, the 'law of physical causality' works behind life. And, a non-reductionist reminds us of the existence of modern mysticism:

> In the past, the failure of embryogenesis to yield to the Cartesian program of

> cut and paste has led to an alternative of an obscurantist holism or, worse, a belief in non-material life force.

Lewontin also teaches me to some extent about the intellectual barriers that prevent a life scientist from being reductionist:

> Yet a study of all or any significant part of the interconnections and coupled oscillations of our millions of brain cells seems a task that would consume the three billion years left before the sun becomes a red giant and fries all of life. We must face the possibility that we will *never* understand the organization of the central nervous system at any but the most superficial level.

If he believes that between theoretical reducibility and practical irreducibility there is a gap wider than three billion years, then being a reductionist simply means being an idealist, or a dreamer, whereas a 'theoretical' reductionist should be a perfectly legitimate status.

In physics, some deep-rooted difficulties pose grave problems that are not just of the nature of practicability. One of them is whether thermal phenomena can be reduced to the random movement of a swarm of molecules. While working physicists — theoretical as well as experimental — all answer this question affirmatively and do their research accordingly, some philosophically sensitive physicists are aware that for a long time theoretical physics has been pestered by a contradiction between two rigorously mathematical theories: the Boltzmann H-Theorem and the Poincaré-Zermelo Cycle-Theorem.

What is clear to everyone is that before using up the remaining three billion years, *homo sapiens* will make some more progress, both materially and intellectually, and hopefully also morally.

6. See Sections 2 and 3 of Essay One (Chapter 2).

With the perspective of three-stage formative development of mathematical mechanics, we can see clearly the resolution of puzzles such as 'why Archimedes didn't set off a scientific revolution', or 'why did Archimedes stop short, 1800 years before Galileo'. (See N. Sivin, 'Why the Scientific Revolution Did Not Take Place in China — Or Didn't It?' *Chinese Science, 5,* 1982; pp. 61–2.) Although 'from Archimedes to Galileo and Kepler', in terms of scientific progress, the time duration did not have to be as long as 1,800 years, there was a big intellectual barrier to be overcome, and Archimedes could not accomplish single-handedly all that was necessary for the fulfilment of a 'mechanical revolution'.

But philosophers heralded the 'mechanical revolution': without mathematics, F. Bacon advocated inductive empiricism; with mathematics, R. Descartes demonstrated the tremendous power of a Euclidean-type methodology. In the early seventeenth-century Western scientists practised sciences in many directions. However, only the mathematical summary of mechanical phenomena could reach a wonderful combination of comprehensiveness, exactness and clarity, which in turn impressed *philosophes* of the following generations. Incidentally, the full blossoming of intellectual creativity in seventeenth and eighteenth-century Western Europe accentuates Archimedes' position as a relatively isolated genius of his time, who could not act as the sole prime mover of a great socio-intellectual movement, which is commensurate to the great Scientific Revolution. In view of the Sino-European difference in intellectual activation, and specifically in mechanical development, what several seventeenth-century Chinese astronomers did, by following Jesuit instruction in Western planetary and calendrical astronomy, can hardly be dubbed a 'scientific revolution'. Traditional China was weak in exact science, and was particularly weak in mechanics. An analysis reveals that along the three stage development, indigenous Chinese mechanics had only attained the first half stage.

7. See Section 3 of Essay One (Chapter 2).

In a recent article, 'Rivers of Science', in the book *China, 7000 Years of Discovery* (1982 by the China Science and Technology Palace Preparatory Committee, Beijing and the Ontario Science Center, Toronto) Needham again addresses a number of basic issues, including, in my terminology, the problem of the criterion of scientific 'modernity'. He writes:

> . . . why had modern science developed in Europe alone? It happened in the time of Galileo, the great astronomer. Leonardo da Vinci, though a genius, did not belong to the time of modern science. One might say also that Tycho Brahe did not, though Kepler perhaps did.

Owing to the original meaning of 'modern', or 'modernity', seeking a chronological criterion for 'scientific modernity' has been a straightforward pursuit. In his previous writings (for example, *The Grand Titration*, pp. 14–15) Needham also shows his emphasis on the time factor. From this quote of his we see that his consistent criterion is Europecentric as well as heavily chronological. (See Section 3 of Essay One, especially the quotation in note 13 of Essay One.)

8. See Section 13 of Essay One.

9. See Section 2 of Essay One, particularly the quote identified by note 4.

10. The 'law of physical causality', which has already appeared in note 5, is explained in Section 5 of Essay Three.

11. In fact year 1922 also saw the publication of Hu Shi's *The Development of the Logical Methods in Ancient China* (The Oriental Book Company, Shanghai, 1922). Although the term 'science' is rarely mentioned in this original study, it is none the less an important book on China's scientific stagnation. After discoursing on the logical methods of various schools of pre-Qin China, Hu Shi concluded his book by translating a two page article from Sima Qian's *Records of a Historian* (first century BC), describing the First Emperor of Qin's brutal suppression of scholarship and learned men. His message is clear.

In the recent issue of the *Journal of Dialectics of Nature*, which contains the special section of inquiry into the causes of China's backwardness (note 2 above), there is a paper by He Xiu, 'On the Academic Disparity between China and the West'. The paper traces China's inability 'to nurture a new type of scientific and technological system' to the difference between pre-Qin Chinese philosophers and Greek philosophers in their aims, attitudes, and methodologies. But the persuasiveness of this analysis is much weakened by Hu Shi's 1922 book, which displays a remarkable richness of logical study in pre-Qin China.

Although Feng You-lan's approach is philosophical, He Xiu's is cultural, and mine is political, we all accept the impossibility of a spontaneous Scientific Revolution in traditional China. Hu Shi tried to adopt a political approach, but he was rather inconsistent on the question of China's stagnation (see note 20 below). As early as 1915, Ren Hong-jun published in *Ke-xue (Science)*, vol. I, 'Shuo Zhong-guo wu Ke-xue zhi Yuan-yin', (Remarks on the Causes of China's Lack of Science). Ren identifies the major cause to the failure to develop the inductive method. Therefore his interpretation may be called methodological.

It seems that the year 1922 represents a highlight in the confrontation of scientific importation and traditional thoughts in China. In that year one of the 'most outstanding 20th century Chinese Buddhist thinkers claimed that Buddhism "included science" '. (See Wing-tsit Chan, 'Neo-Confucianism and Chinese Scientific Thought', *Philosophy East and West*, vol. 6, 1957, p. 311.)

12. B. Russell, *The Problem of China* (G. Allen & Unwin, London, 1922), p. 10.

13. Ibid., p. 194.

14. In order to understand our general argument (in fact a mechanistic or

reductionist stand), it may be useful to quote another contrast in this connection. Russell wrote in his 1922 book, *The Problem of China*, pp. 81–2:

> The real problem for the Chinese intellectuals is to acquire Western knowledge without acquiring the mechanistic outlook.
>
> Perhaps it is not clear what I mean by 'the mechanistic outlook'. . . . What I mean is the habit of regarding mankind as raw material, to be moulded by our scientific manipulation into whatever form may happen to suit our fancy.

Now Lin Yu-tang in his essay, 'Han Fei As a Cure for Modern China' (Hu Shi & Lin Yu-tang, *China's Own Critic*, Paragon Book Reprint Corp., New York, 1969, pp. 94, 93), wrote sometime before 1933:

> What interests me most is that here we have a mechanistic conception of government by law pushed almost to the extremes, a theory that would be more typical of the German mind than of the Chinese way of thinking.

By 'a mechanistic conception of government' Lin meant that

> in contrast to the Confucian dictum that 'ceremony should not be applied to the plebians, and punishment should not be extended to the lords', we have here a legalist who says that we should have a law that doesn't fawn upon the mighty, a rule that should be applied rigidly, so that wherever the law applies, the clever will submit and the powerful will not protest, that the nobility should not be exempted from punishment and rewards should not go over the heads of the humble.

Indeed we do not quite follow Russell since he puts 'our scientific manipulation' and 'our fancy' together. I think human laws are something between science and fancy — neither purely scientific, nor purely fanciful. Consequently, I think that in the above-quoted passages, we may understand that Russell and Lin were more or less talking about the same thing, about what philosophical outlook and mental attitude should the twentieth century Chinese adopt — in order to survive and prosper in a rapidly changing world, or to preserve traditional, 'precious' Chinese virtues (cf. note 12 above).

15. Yu Lan Fung (Feng You-lan), 'Why China Has No Science', *The International Journal of Ethics*, vol. 32, no. 3 (April 1922), p. 237.

16. I certainly appreciate the great significance of this summary *per se*. Besides philosophical and methodological differences between the Needham school and myself, however, I can add the following metaphorical explanations: when one is seriously ill one needs a doctor, that is to say one needs a truthful diagnosis in the first place. The patient does not need a 'historian-prophet' who is preoccupied in lecturing about how healthy the patient had been and how strong the patient will be.

My other metaphorical statement is: excessive glorification of my ancestors is in fact an insult to me.

17. (1) By macro-history I mean that the questions with which we want to deal and the factors behind these questions have an enduring character. (Thanks to the existence of factors of such a character, it is meaningful to discuss historical questions which cover a long span of time.) In a sense, my faith in macro-historical reasoning is based on the analogy that in engineering design, thermodynamical calculations (which are essentially a 'black box' methodology, see Sections 7 and 9 of Essay Three), are important, and in many cases necessary. Macro-historical reasoning helps us to judge statements like 'until the middle of the 17th century Chinese and European scientific theories were almost on a par', or to answer questions such as whether traditional China had rational and theoretical inquiries in mechanical science.

(2) Once with a young colleague of mine, Mary, I had an interesting conversation which is related directly to my idea of macro-historical reasoning.

After reading some of my writings, Mary asked me: 'Macro-history and macro-historical reasoning are your favorite phrases. What do you exactly mean by them?'

I responded: 'Yes, I really should have explained them . . . May I tell you one of my cultural shocks so that you can understand my macro-historical reasoning?'

Mary: 'Sure. I'd like to know all your cultural shocks.'

I: 'But let's keep to this particular one at the moment.'

Mary: 'I agree. Go ahead please.'

I: 'You see, until now I have never been to a Christian wedding ceremony, but I've seen it many times on TV or in a film, and now I can almost repeat what the priest and the new couple are going to say. To tell you the truth, before I came to the West, I was not aware that marriage for Westerners was so religious, or completely religious, a commitment. My understanding was in fact typical among the Chinese, because China can now be rightly described as an agnostic nation.'

Mary:'Nothing religious in a Chinese wedding?'

I: 'In the past a wedding asked the Heaven and the Earth to witness, but that was all about its spiritual aspect.'

Mary: 'But is that your cultural shock — just listening to what people say at a Christian wedding ceremony?'

I: 'No, that does not comprise a cultural shock yet. It was when I put the wedding oath and the high divorce rate together I went through a kind of cultural shock, and I said to myself, "Here I have pinpointed some Western hypocrisy!"'

Mary: 'But you don't really mean that the wedding oath is hypocritical, do you?'

I: 'But I do. I certainly admit that most of the new couples love each other passionately at the very moment, and they express their vow sincerely, by no means hypocritically. But, it is when we put the two ends together — when we juxtapose the solemnity of the ceremony and the high rate of divorce that "till death do us part" sounds hypocritical.'

Mary: 'But not to me.'

I: 'But you can understand that it does so to me, to a pair of Chinese ears. A Chinese couple at marriage do not say that, while Chinese marriages are statistically stable. A Western couple do say that, while there is about more than one-third probability that they will part before death, and in many cases because of "wealth, poverty, health, and sickness". One may write a big volume on a Sino-Western comparative study of marriage-divorce, yet I believe my observation would retain its undeniable validity.'

Mary: 'Do you mean a kind of "rational reconstruction"?'

I: 'No. I think macro-historical reasoning differs from rational reconstruction. One essential ingredient of history is being faithful to facts, as much as possible, and so for macro-history. My statement about present-day Western marriage is not only true to a pair of Chinese ears, but it is true objectively. If the current divorce rate is 45 per cent, then the measure of the sincerity of the wedding oath is 55 per cent, and the measure of its hypocrisy is 45 per cent.'

18. I certainly do not deny Needham's love of China, China's past, present, and future. But in reading his essay 'The Past in China's Present' — from the institutions and social practices that he praised — I was surprised by his gross misunderstanding of the basic aspiration of the people in present-day China. J. Needham, 'The Past in China's Present' (Far East Reprint; reprint from Arts and Science in China, London; 1959 or later). In a nutshell, Needham tells us that the Chinese people would be very happy in an ideal, or the best, traditional Chinese society, and that under the rule of Mao and his party, China is, or approaches, that ideal society. The actual Chinese development that followed Needham's eulogy of communist dictatorship, the communes, the Great Leap Forward and so on

demonstrates, I think, two basic Chinese traditionalisms: one is into what an extreme absurdity the Chinese autocracy can lead the nation, and the other is the extreme hatred the people harbour against the Chinese autocracy. That this dreadful situation is still a living reality in China evinces that the Maoist one-party dictatorship represents at once the worst and the most effective Chinese autocracy.

19. In an 1982 article, 'Rivers of Science' (note 7 above), Needham addresses the stagnation problem this way:

> Many people often realize some of the great things that were done in China in ancient and medieval times, but after that they talk of a 'stagnation' which overcame Chinese culture. We do not believe in this at all; what we say is that the course of discovery and invention in China went on slowly all through the ages . . .

20. A typical passage of the theory that 'hardware decides' belongs to Hu Shi. Hu Shi, *The Chinese Renaissance* (The University of Chicago Press, 1934), p. 63:

> The difference between the Eastern and Western civilizations is primarily a difference in the tools used. The West has during the last two hundred years moved far ahead of the East merely because certain Western nations have been able to devise new tools for the conquest of nature and for the multiplication of the power to do work. The East, whence have come a number of the epoch-making tools of ancient civilization, has failed to carry on that great tradition and is left behind in the stage of manual labor while the Western world has long entered the age of steam and electricity.

A slogan painted on the wall of a hall in Chicago's Museum of Science and Industry provides another example: 'Tools make America great.'

My theorem that 'software decides' came from pondering over Chinese history: traditional China invented important hardware such as paper, printing, gunpowder, and the magnetic compass, yet it was the persistent Chinese software, the unified imperial institution combined with strict ideological control, that played the decisive role.

21. The Institute of the History of Natural Sciences, Academic Sinica (eds), *Zhong-guo Gu-dai Ke-ji Cheng-jiu (The Scientific and Technological Achivements in China's Past)* (Beijing, 1978; 1980 reprint). In Essay One (Chapter 2) this often-quoted book is referred to by its English acronym, STACP.

22. This formulation belongs to Professor W. Monter.

23. Edwin Reischauer, *The Japanese* (Harvard University Press, 1977), p. 52.

24. At the Buffalo meeting I still used the phrase the 'principle of the degree of intellectual mobilization'. After discussion with Professor Joravsky, I have now decided on the 'principle of the degree of intellectual activation and creativity'. When developing a theory on the growth of modern science I first used the Maoist phrase 'mass-movement', by which I meant an inheritable 'mass-movement' among intellectual elites. After that I used the Maoist term 'mobilisation'. (I am an eclectic man.) But it was not implied that there were necessarily conscious efforts on the part of political or spiritual authorities to mobilise intellectual elites to engage in intellectual activities in various fields. I meant that the social environment itself 'did' the 'mobilisation' — by being conducive, or neutral, or inimical. Therefore, 'mobilisation' is not an accurate term; 'activation' is good, and 'creativity' is necessary.

25. Needham, *Science and Civilization in China*, vol. II, p. 294.

26. Hu Shi (note 20 above), p. 45.

27. Ibid., pp. 45–6:

> But all these great movements which rightly deserve the term of 'renaissances', suffered from one common defect, namely, the absence of a conscious recognition of their historical mission. There was no conscious effort nor articulate interpretation: all of them were natural developments of historical tendencies and were easily overpowered or swept away by the conservative force of tradition against which they had only dimly and unconsciously combated . . . they brought in new patterns, but never completely dethroned the old, which continued to co-exist with them and in time absorbed them.

28. *Da Xue (The Great Learning)*, one of the standard Neo-Confucian texts, which group formed the required reading for every Chinese student during the six centuries before 1905, begins this way: 'The law of the Great Learning consists in developing and making clear the luminous principle of reason which we have received from Heaven . . .'

2 THE PHENOMENOLOGY OF CHINESE STAGNATION IN PHYSICAL SCIENCES

Bing-dong san-chi fei yi-ri zhi han.
[It takes more than one day's cold to freeze the river three feet deep.]

— a Chinese folk saying

1 Stagnation Shown by Comparisons

Mainly by way of comparing the macro-historical features of the development of physical sciences in China and in Western Europe, this chapter intends to show the long delay on the part of the former. Since the 'mechanical revolution' played a leading role in the Scientific Revolution between 1500 and 1700 in Western Europe, it is crucial to analyse the status of China's mechanical development in order to understand the scientific potential of traditional China. This statement is based on the assumption that the 'mechanical revolution' played a role of universal character in the Scientific Revolution.

Since objective verifiability and conformity to rigorous logic are inherent conditions of science, people generally recognise one science — the totality of all systems of knowledge and methods that are experimentally verifiable or at least, logically supportable to a relatively satisfactory extent. Did the historic growth of modern science have to follow a single pattern of development? I am aware that this is a highly controversial question. The theme of this paper is 'phenomenology' — the description of phenomena and the presentation of contrasting phenomena. In this way, I hope I can stick to a modest, matter-of-fact approach. However, owing to reasons that will soon become clear, we cannot avoid a theoretical discourse on the question of the 'path of modern science', or more specifically, the 'path of physics'. Interestingly enough, the discussion turns out to be very meaningful, because it leads us to the assertion of a 'criterion of scientific modernity'.

Although this chapter bears a general title, it certainly does not claim to be either a full-fledged, or an original, research report.[1]

A considerable part of my material came from Joseph Needham's famous series of work. In fact, my foregoing theoretical question was prompted by Needham's hypothesis that physics might have developed differently. Needham arrived at this conclusion because he believed (a) that magnetism in traditional China was far ahead of the contemporary world, and (b) that Newtonian 'billiard-ball' physics was not a necessary stage in the building of physics. I believe to any physicist the second statement presents a challenge and, as a Chinese physicist, I deem it my duty to rise to that challenge. Now I also believe that the convincing refutation of both statements[2] naturally answers our theoretical question in the realm of physics, and owing to the role physics played in modern science, very probably also for science as a whole. Consequently, this chapter combines theoretical discussion with factual presentation to argue that, phenomenologically, there was a Chinese stagnation; with a theoretical analysis, it can be shown at what a (rudimentary) level Chinese physical science stayed, or changed very slowly.

2 The Scientific Revolution Led by a 'Mechanical Revolution'

Historically modern science grew only in Western Europe. That is to say, there was only one Scientific Revolution. It was in the first few decades of the seventeenth century, in different parts of Western Europe, in different fields of scholarship, that a number of intellectual achievements triggered the irresistible and irreversible (unlike the situation in ancient Greece) growth of modern science: Tycho Brahe's accumulation of accurate astronomical data and his modified Copernican system (before and around 1596), Gilbertian magnetism (1600), Galilean theorems of various forms of a terrestrial falling body (1602, 1603, 1638), Keplerian planetary theorems (1609, 1619), Baconian empiricism and his passionate advocation of the organised scientific cause (between 1605 and 1627), Cartesian mathematisation and axiomatisation(1637) and Harveian anatomy (1638). By the 1660s, the first scientific societies and journals had appeared in England, France, Italy, and Germany, which may be viewed as the official institutions of modern science. At the same time, inspired by an emergent scientific enthusiasm, and modelling Cartesian methodology,[3] Newton achieved a mathematical synthesis of the positive and best results of Kepler and Galileo, an achievement which has since been recognised as the culmination of the Scientific Revolution as well as the beginning of modern physical science.

Can we hypothetically imagine a different pattern to the rise of modern science? A different path of development would mean physical science not being first 'modernised' in a mechanical synthesis of basic planetary and terrestrial movements. Would that be possible? Needham would answer this question affirmatively. On the first page of his volume on Chinese physics he wrote:[4]

> Most important in this Section will be, of course, the development of knowledge about magnetism, in particular the discovery and exploitation of the directive property of the lodestone. The Chinese were so much in advance of the Western world in this matter that we might almost venture the speculation that if the social conditions had been favorable for the development of modern science, the Chinese might have pushed ahead first in the study of magnetism and electricity, passing to field physics without going through the stage of 'billiard-ball' physics. Had the Renaissance been Chinese and not European, the whole sequence of discoveries would probably have been different.

But 'passing to field physics without going through the stage of "billiard-ball" physics' is a misunderstanding of the inherent logic of physics, which, as I want to show below, is causally exhibited in the actual history of physics. Michael Faraday, William Thomson (Lord Kelvin), and James Maxwell initiated and mathematised the concept of a field of force in the process of establishing electromagnetism. Their major aim was to get rid of 'action at a distance' and provide electric and magnetic actions with the tangibility of 'billiard-ball interactions';[5] their basic assumption was that the force in a 'field of force' is the same Newtonian force which governs billiard-ball interactions. At present, both mechanics and electromagnetism explain physical phenomena in their respective areas satisfactorily; in addition, the two areas are logically unified — with the unified recognition of mechanical force as the manifestation of momentum and energy exchanges, either in ponderable bodies or in fields. No physicist today would venture to turn the system logically upside down, so that electromagnetism provides basic concepts for mechanics. We can imagine constructing a multistorey building by starting from the top floor, but we can not imagine people accepting this *ad hoc* procedure as an economic, logical, or historical way of doing things. Mechanics serves as the basis of physical science because it deals with the simplest and immediately applicable natural phenomena. In other words, both human needs and human intellectual curiosity demand a matured mechanics first. Besides, for founding a mathematised system, mechanics is also the easiest branch of physics. (That is why it naturally provides basic concepts for other branches of physics, whereas the reverse would be awkward, if not impossible.) Therefore, if we compare mechanics and electromagnetism we arrive at the following form as set out in Table 2.1.

Table 2.1:

	Mechanics	Electromagnetism
natural phenomena the subject accounts for:	movements in daily life and in general astronomical observations	comparatively special
the formulation of the mathematised axioms of the subject:	comparatively easy; requires no advanced mathematics	difficult; requires partial differential equations
the mutual relation of the two compared subjects:	naturally provides basic concepts for other branches of physics	almost inconceivable that an 'initially-established' electromagnetism would provide basic concepts for a 'subsequently-established' mechanics

Evidently this comparison can also be carried out between mechanics and other branches of physics and one still reaches the same conclusion: mechanics must lead, or mechanics naturally serves as the logical basis. Admittedly the studies of light, sound, and heat started independently of mechanics. Yet it did not take much time for the modern era — since the appearance of the *Principia* — to recognise that the different fields of physical phenomena were interrelated; in fact the rapidity with which physicists arrived at this recognition justifies our concept of a 'physical revolution'.

Moreover, intensive development of analytic and celestial dynamics in the two centuries following Newton reinforces our argument that 'classical mechanics' was the only possible 'leader' in the 'physical revolution'. Eighteenth- and nineteenth-century mathematical geniuses like d'Alembert, Euler, Lagrange, Laplace, Gauss, Jacobi, Hamilton and others, were all keenly interested in mechanics. They further developed the *Principia*, although none of them transformed it. It was not until 1905 that Einstein's special relativity revolutionised the traditional concepts of space and time and established a new system of mechanics. What is the moral of this macro-historical consideration? In view of the conceptual (though not mathematical) sophistication of special relativity and its reliance on the maturity of electrodynamics, the conclusion can only be that 'Newtonian' mechanics was the only system that human intellect could first arrive at to comprehend the mechanical movements around us.[6] The methodology of exact science — accurate and verifiable formulations, observation, experimentation, rigorous logic, and mathematisation — also appeared spontaneously and sporadically in areas other than terrestrial mechanics and planetary astronomy (we will see below that in traditional China there was the brilliant case of experimental and theoretical studies of the musical scale), yet the scientific breakthrough had to happen in the way it did.

If this is a conceptual refutation of the Needhamite speculation that field physics might have risen without going through the stage of 'billiard-ball' physics, we can also provide a simple historical refutation. We just make a comparison of the best results of traditional China's magnetism and Gilbert's *De magnete*. If we find that Gilbertian magnetism had summarised everything that traditional China knew about the lodestone until 1600, and if Gilbert's treatments were more systematic and more science-oriented, then the 200 year discrepancy between the founding of the basis of mathematical mechanics and electromagnetism shows nothing short of the total collapse of the Needhamite hypothesis — because the Renaissance was a European movement and Gilbert was a late Renaissance man. In fact, some major developments of electromagnetism that extended about three centuries occurred nowhere other than in England, developments which can be conveniently schemed in a double series:

W. Gilbert	I. Newton	H. Cavendish	H. Davy	M. Faraday	Lord Kelvin	J. Maxwell
magnetism	mathematical mechanics	electricity (no direct historic contribution)	(among others) electro-chemistry	(mainly experimental) electro-magnetism	(saliently engineering) electro-magnetism	mathematical electro-magnetism

The last part of this chapter compares Chinese magnetism and Gilbertian magnetism more systematically, and we will better understand both China's scientific stagnation and Gilbert's historic position in Western scientific development.

It is obvious that the Gilbertian magnetism was more complete, more scientific, and more intellectually stimulative. Therefore, logically Needham's hypothesis infers that England, or Western Europe, should have first developed a field physics based on electromagnetism. The actual history of physics as we have just pointed out, directly gainsays this theory. The positive moral that results from this confrontation between historic facts and a historical hypothesis is that 'the whole sequence of discoveries' of physics could *not* have been 'entirely different'; on the contrary, 'the whole sequence of discoveries' of physics could have only developed by and large as it did in actual history. Indeed, we even don't have to worry about historic fortuity in the discovery of facts. The establishment of mathematical electromagnetism was a mature process that progressively correlated phenomena with theories. As early as in Gilbert's study of magnetism, which was conducted in the last two decades of the sixteenth century, scientific fact-finding showed features of a systematic laboratory work.

I am aware that some historians of science may not like the conclusion I have arrived at. Since the Second World War, one intellectual trend was not happy about the fact that of all civilisations around the world, *only* that of Western Europe produced *modern* science. Now the implication of my above analysis is that European scientific development, especially in the general pattern of (though not as an equation)

enthusiasm in experimental and exact sciences		a unified (classical) physics
daily life locomotions	→ mechanics →	mathematics
planetary movements		confidence in theorising exact sciences, and in enhanced rationalism,
mathematics		

also serves as an 'objective' yardstick to 'measure' the historic development of modern science in other civilisations. At the time when the 'mechanical revolution' occurred, science also made progress in other fields, but mechanics undoubtedly held a pivotal position in the whole picture. This was history in Western Europe. Should we gauge the historic development of other civilisations by the history of Western Europe? Taken literally, this question sounds absurd, even outrageous. But we have to be aware that we are talking about the history of *science*. As I have emphasised at the beginning, there are no different and parallel systems of science. By and large we do not have different systems of physics. The same gravity acts upon a Chinese, an Englishman, and an African. By and large, peoples on different continents were concerned about the same range of natural phenomena and technical artefacts: basic celestial movements, the lever, a buoyant or an immersed body, a piece of lodestone, the use of wind and water power, wheeled vehicles, etc. A few people from time to time were engaged in rational inquiries about these phenomena, and after very uneven and tortuous development, finally succeeded in unifying them theoretically. Other peoples were not, or unable to be, interested in such intellectual engagements. However, on the one hand, these peoples could not offer alternative systems which were equally comprehensive, exact, and useful; on the other hand, when these peoples finally became attracted to, or were forced to accept, modern science, by and large, they found it convincing and beautiful. Therefore, I do not see any more unpleasantness in the existence of a historical criterion for scientific 'modernity' than in any specific Western priority in scientific discoveries.[7] To summarise, in relation to other branches of physics, mechanics was historically as well as logically inexorable in two aspects: one, it was the first field to be axiomatised, or to attain the status of a theoretical synthesis; two, it acted as the primary field, on the basis of which theoretical understanding of other branches of physics developed. (Optics was the remotest. Mathematically, it was related to mechanics via electromagnetism; experimentally, when the pressure of light was detected, both in the late nineteenth century.)

3 Some Applications of the Criterion

By this criterion, it is *not* Taoism that represents the foremost scientific thinking in Chinese thought — although we may agree with Needham that we are indebted to Taoists for 'the beginnings of chemistry, mineralogy, botany, zoology and pharmaceutics in

East Asia'.[8] On the basis of his faith in the organicist philosophy, Needham asserts to the contrary. The reasons why we have to dismiss Taoists as the leading Chinese scientists are provided by Needham himself: because Taoists 'mistrusted reason and logic', and because, as Wang Chong criticised them in the *Lun Heng* (AD 83), 'The Taoist School . . . does not know how to substantiate its cause by evidence.'[9] Our stand is strengthened by pointing to an analogy: no one would agree that alchemy should logically be put prior to Ptolemaic astronomy in the rank of contributors to the growth of modern science. By our criterion, Mohism and the thinking of the Logicians were more scientific than all others. The general reason is perspicaciously stated in an observation on Mohist optics and mechanics by Graham and Sivin:[10]

> To an extent quite unusual in early Chinese science we find ourselves in a landscape of geometrically visualised figures, very far from the metaphysics of yin and yang and the Five Phases, of numerology, time cycles, and resonances. The same is true of the propositions on mechanics.

By our criterion the following quotation is a good passage in experimental physics. It is a passage from the chapter 'Things of the Same Genus Energize Each Other' in *Chun Qiu Fan Lu*[11] by Dong Zhong-shu, a powerful prime minister in the former Han Dynasty who 'advocated Confucianism only, and abolished the One Hundred Schools':

> . . . if notes correspond, they resonate. The experimental proof of this is extraordinarily clear. Try tuning musical instruments. The *gong* note or the *shang* note struck upon one lute will be answered by the *gong* or the *shang* notes from other stringed instruments. They sound by themselves.

Resonance is admittedly not a simple phenomenon, and acoustics is almost as poor a candidate for the first field of modernised physics as electromagnetism. Yet if systematic experimentation had been carried out around descriptions such as this, early scientists might have familiarised themselves with the scientific methodology, and certain generations of their descendants might finally have attacked the right points in the front, thus paving the way to successful breakthroughs. But I do not mean to imply that this particular piece of Dong Zhong-shu's writing is very encouraging for experimental endeavours. From the context it is clear that the author is interested mainly in philosophy. He uses a number of examples and analogies, of which acoustic resonance is one. All the others are very unscientific, for example, 'if (two) *qi* are similar, they will coalesce', or lovely things summon others among the class of lovely things'. (This pre-modern idea also appeared in the West, namely the dichotomy of 'sympathy/antipathy', that is, 'like attracts like', in fact a reversal of the bi-elementalism of the Yin and Yang. In Europe it was Gilbert who replaced this archaic conception with the concept of physical forces, an innovation that was to contribute significantly to the fruition of modern mechanics.)

Our objective criterion of scientific 'modernity' is not chronological. There had already been some vital scientific seeds in Hellenic civilisation. Thumbing through the books of Euclid, Archimedes, Apollonius, and other ancient mathematicians — rendered into modern notations — one cannot help being struck by their scientific modernity and unvarnished professionalism. Yet when we try hard to make sense of the propositions on mechanics and geometrical optics in the roughly contemporaneous Mohist Canons, we find ourselves more often than not in a helpless quandary.[12]

The evaluation that an idea about nature was historically helpful, or progressive, or obstructive, or backward, may raise difficult questions. Now our objective criterion of scientific modernity may shed some light on the discussion of such questions. After telling of a fifth-century inconclusive discourse on the fundamental idea of the *Yi Jing (The Book of Change*, that might have originated as early as the eighth century BC), Needham comments:[13]

> Up to the present point the consideration given to the theories of the Five Elements and the Two Forces has indicated that they were a help rather than a hindrance to the development of scientific ideas in Chinese civilization. Not until the 17th century, when the four Aristotelian elements were finally discarded in Europe, did these theories confer upon Chinese thought any degree of backwardness as compared with the world-picture of occidentals.

The Chinese Yin and Yang or the four Aristotelian elements were progressive ideas when they were first put forth because, I think, they represent ancient thinkers' nascent theoretical efforts to explain phenomena by a small group of essentials. But I would assign the time of their transition from being progressive to being backward much earlier, because (1) they were vaguely defined, (2) they made only *non-sequitur* explanations, (3) they predicted nothing precise, and (4) they were not subject to testing. Moreover, I would not say that a Chinese concept became backward by the fact that in Western Europe an equivalent concept was replaced by a more scientific idea. As is made clear in the above quotation, Needham *chronologically* measures Chinese development with a *Western* yardstick. I measure Chinese development differently, that is, against the concrete contents along the inexorable growth of modern science, particularly the sequence of discoveries and intellectual creations that led to classical mechanics. The yardstick that I use is also Western because, as I have shown above, in matters of physical sciences, historical sequence on the one hand and inherent logic on the other conform to each other.

Given the Western inheritance of the Greek legacy as early as the translation movement in the twelfth century (alongside a continuous, though inconspicuous, lineage in the forms of encyclopaedism and church school curriculum), it is evident that the following judgement by Needham is not correct:[14] 'Until the middle of the 17th century Chinese and European scientific theories were almost on a par, and only thereafter did European thought begin to move ahead so rapidly.' Indeed this statement poses an unsolvable riddle when we realise the sharp contrast that in Europe the middle of the seventeenth century saw the birth of mathematised and axiomatised mechanics, whereas the best intellectual representatives in China, limited by their politico-academic environment, could not even appreciate this form of mechanics until two and a half centuries later.

Sometimes the following question is credited as 'Needham's discovery':[15] How was China able to attain an early scientific and technological superiority over the West, so that until the Renaissance it had more to give than to receive? But because of the rudimentariness of Chinese mechanics and China's lack of the notion of mathematical rigour, the assumption behind this question is technologically partly true and scientifically untrue. With regard to the most vital branch of exact science, not only would traditional China give nothing, but it was for centuries not even in a position to receive: from the thirteenth century to the end of the nineteenth century, time and again China declined to import and digest the best of Hellenic science.[16]

Another question compares China with one particular nation in Western Europe, specifically, with France. Sivin writes:[17]

> Only habit distracts us from recognising that the introduction of modern cosmology or mechanics into France and their importation into China, if they are not precisely equivalent historical problems, at least share many dimensions. It is equally to the point — or equally beside it — to ask why France was incapable of developing modern science solely out of her own inner resources as to pose the same question about China.

By 'not precisely equivalent historical problems' Sivin may mean that France has always been an inseparable, and indeed central, part of European culture. 'Why China persistently showed incompetence in accepting more efficient institutions and more dynamic ideas' and 'why Shanghai developed differently from an average interior Chinese city' are obviously questions on two different levels. Besides this incongruity, the specific juxtaposition of China and France misses the mark because we know that since the late Middle Ages the University of Paris had been much involved in the parallel and opposed developments of Christian theology, Aristotelianism, and positive knowledge. Particularly on matters of 'modern cosmology or mechanics', in the fourteenth century Jean Buridan and other scholars were actively engaged in the discussion of the basic conception of mechanics. Historians of science generally recognise that:[18] 'It is clear that Buridan's impetus theory marked a significant step toward the dynamics of Galileo and Newton, and an important stage in the gradual dissolution of Aristotelian physics and cosmology.' Although I emphasise modern science as the result of collective efforts of Western Europe as a whole, we can evaluate France's capability (not incapability) of developing modern exact science by looking at some obvious facts. In the early seventeenth century, just before the Newtonian culmination of the Scientific Revolution, France produced Descartes (1596–1650), Fermat (1601–65), Pascal (1623–62), and others (see note 3). One may even say — with perhaps too much mathematical profundity and not too distant an analogy — that Fermat anticipated Newton's mechanical laws by his 'principle of least time'. Then, during and since the eighteenth century, in developing the

Principia, France again produced Voltaire (a passionate champion of Newtonian science, just as a century before Bacon had been for science in general), d'Alembert, Laplace, Lagrange, Leverrier, and others.

Sivin thinks his paper, 'Copernicus in China', illustrates the general process of 'how one culture accepts and rejects the scientific ideas of another'. I believe this is a useful inquiry when studying, for instance, China and Japan, or comparing the two. But it may not be an appropriate question when making a comparison between China and France, because the latter can hardly be alienated from the source of those vital ideas of exact sciences, whereas China's incapability of developing modern science has been so conspicuous that, even with conscious and official importation, the state of non-development nevertheless dragged on and on.

By our criterion, it is not the verifiability of a specific statement that decides its acceptability. At any stage of modern science there have been many more unverifiable scientific statements than true ones. Therefore, Dong Zhong-shu's acoustic description, or Aristotle's mistaken 'law of falling bodies' were scientifically progressive, because they were clearly formulated, they predicted experimental results (in one case, quantitatively), and they were subject to test. The scientific significance of Aristotle's 'law of falling bodies' was fully revealed by the unsubstantiated episode of Galileo's public disproof of it. Whether or not the Leaning Tower Experiment at Pisa was actually carried out, terrestrial mechanics made decisive progress in the hands of Galileo — partly because he conscientiously evaluated and tested Aristotle's scientific, though unverifiable, hypothesis.

By our criterion, the best science in the Hellenic tradition was Euclidean geometry and Archimedean mechanics. The former provided an axiomatic model and rigorous logic, and the latter offered mathematisation of simple mechanical phenomena. The nearest comparable works among early Chinese thinkers to Greek exact science were those of the Mohists (Mo Ti lived within the period 479 to 381 BC) and of the Logicians (Hui Shi lived during the fourth century BC, and Gong-sun Long, in the first half of the third century BC). The Mohists, a group of compassionate artisan-philosophers of the dynamic late Warring States period, had started to formulate logic, economics, mechanical engineering, geometrical optics, and mechanics in their philosophical argumentations. However, it was a great loss to Chinese science that in the long history of traditional China, Mohism, as a staunch rival school of Confucianism, was almost completely left in oblivion. The Logicians fared no better.

4 Chinese 'Theorems' of Buoyancy

What could be collected as the Chinese equivalent to Archimedes' principle of buoyancy were one Han episode (which occurred just after AD 200) and an earlier Han passage.[19] The episode of 'Cao Chong weighing an elephant' has been a popular Chinese story. Physically it assumes that a certain weight in a boat sinks the boat to a certain depth. The story itself is certainly a far cry from Archimedes' principle. Even when the simple principle involved is mathematised in the above form, it is still far from being equivalent. The problem is that in China's long history, although the story praising Cao Chong's cleverness was told again and again, no scholar was interested in generalising from it.

Needham thinks that the question of whether or not Han technicians had mastered the principle embodied in Archimedes' determination of the proportion of gold and silver in Hieron's crown 'depends on the interpretation of an important passage':[20] 'The workers . . . make measures of capacity. They purify (separately) by successive heating samples of metal (presumably copper) and tin, until there is no further loss of weight. Then they weigh them.' (Words in parentheses are apparently inserted by Needham. The first, 'separately', doesn't make sense to me; the second insertion is discussed later.) Biot translated the following passage this way: 'After they have weighed them, they proportionalise (or equalise) them, and after that they measure them.' Needham now points out[21] that Jiang Yong, however, had already suggested at the end of the eighteenth century that the second action, the one translated by Biot as 'proportionalising' or 'equalising', had meant in the Han era 'weighing in water'. By this explanation Needham states, 'What the Han technicians were doing, therefore, was essentially what Archimedes did, namely to ascertain the proportions of an alloy by weighing in water as well as in air.' Then he concludes: 'Such an interpretation makes more sense than Biot's.'

To be frank I am at present not in a position to pursue the archaic meanings of the character 'zhun'. It is understandable that contemporary Chinese do not know that it means 'weighing in water', because in any historic period weighing in water should be a rare operation whereas as a commonly used character today, 'zhun' conveys the meanings of 'standard', 'accuracy', and 'levelling', (and for these meanings it is conceivable that the character contains the 'water radical'). However, even if we spare etymological discussions, I believe we can still judge whether or not the Han technicians were ascertaining the compositional proportion of an alloy, because the sophistication of the procedure was evidently beyond the level of the Han technicians. What I want to show is that we can follow either Biot's translation or Jiang Yong's explanation, but neither will lead us to Needham's conclusion. Many present-day Chinese, including me, would agree with Biot before they learned about Jiang Yong. Now since we are informed about Jiang Yong's interpretation of the character 'zhun', let us follow him and examine Needham's logic.

We should notice that Archimedes could solve his question because, firstly, he was a good mathematician. (Incidentally in order to fulfil the task assigned to him by King Hieron, Archimedes didn't have to know the Archimedes' principle on buoyancy. What he needed to know was the volume of the crown, which could be measured by displaced water. Therefore, taking a bath in a tub was crucial to Archimedes' 'Eureka' indeed! On the other hand, with the aid of his principle, which asserts that the loss of weight of a body in a liquid equals the weight of the liquid which is displaced, the volume of the crown could be easily determined by weighing the crown in and out of water.)

Secondly, Archimedes knew that there were at most two components in the alloy. If there were three or more components, then he could only affirm whether there was adulteration or not, but not ascertain the composition quantitatively. A set of two equations with more than two variables has an infinite set of solutions.

In the above-quoted Chinese passage, obviously it was Needham who inserted '(presumably copper)' behind the word 'metal'. If the Han technicians were interested in the compositional question, they should be more specific about the 'metal'. And if there were more than two 'metals' in an alloy, ascertaining its composition was simply impossible by weighing the resultant weights (in and out of water) or measuring the resultant volume.

The first part of the quotation tells us that those technicians were producing standard capacity (volume). For this purpose, what is the point of knowing the composition of the alloy? From the second part of the quotation, according to Jiang Yong's explanation, what might be inferred is that the Han technicians mastered 'Archimedes' principle' to the extent — still far from an exact and full formulation of the principle — that if the weights weighed the same in the air, and assuming they were of the same alloy composition ('successive heating . . . until there is no further loss of weight'), then they have the same capacity if they also weighed the same in water — consistent to their purpose of making standard capacities.

Apparently I am not the only person to have noticed that Needham is sometimes inclined to overextend his extrapolations. In *American and Chinese* (The Natural History Press, pp. 371–2), Francis L. K. Hsu observed that equal temperament, which we will discuss in Section 13, 'is commonly credited to the European Andreas Werkmeister. But Needham, after considering all sorts of evidence, infers that an independent invention of equal temperament in Europe is most unlikely.' Then Hsu quoted Needham's own words,

> In any case it is fair to say that the European and modern music of the last three centuries may well have been powerfully influenced by a masterpiece of Chinese mathematics, though proof of transmission is not yet available.

In the case of the origin of Leibnizan organicist philosophy, which we will meet more officially in Section 15, Essay Two (Chapter 3) and Section 10, Essay Three (Chapter 4), Needham repeats the same pattern of argument. To advocate his 'Chinese correlativism' he would write, 'if I may whisper it for the first time, Organism' (*Science and Civilization in China V (II)*, p. 281). More often he likes to repeat his logically unfinished arguments again and again. In a recent book, *Revolution in Time* (Harvard University Press, 1983), David Landes writes (pp. 20–1):

> Unfortunately, Needham, in his eagerness to restore the good name of Chinese horology, was not content with what must surely rank among the great detective coups of historical scholarship. He was convinced that there was more to these astronomical waterwheel clocks than the culmination of the hydraulic line of horological development. That would be a dead end, however magnificent. He preferred to see them, rather, as a

> beginning, as forerunners of the European mechanical clock.
>
> Now this is optative history, the story of what should have been. There is not even a bond of logic or method between them. . . .
>
> In any event, neither the systemic differences nor the absence of corroborating evidence discouraged Needham. Using a *post hoc, ergo propter hoc* reasoning, he speculated that rumors of the Su Song clock somehow filtered westward over a period of a hundred or more years and struck a spark . . .

Then on page 23 of his book, Landes helps us:

> To understand this figment, one has to place it in the larger context of Needham's philosophy of history . . . what he calls ecumenical science . . .

Finally let me quote Landes again to conclude this insertional critique of Needham's manner of argument:

> This [the 'ecumenical science'] is a noble credo, but it should not be allowed to get in the way of historical accuracy . . . Joseph Needham is too great a scholar not to be taken seriously, but what does one do when his wishes become affirmation? Carlo Cipolla sees all this as fancy. I agree with him.

We now return to the question of Chinese knowledge on buoyancy. If the Han hydrostatics really attained the maturity of using Archimedes' principle to ascertain the composition of a two-component alloy — in other words, if Needham was correct — then the concept of specific gravity should have been familiar to Han technicians and interested scholars. Yet it seems that written references to specific gravity or density in Chinese history were rare. Needham, however, believes that the 'general idea of specific gravity must have existed from time immemorial'. His reason: 'Mencius (fourth century BC) remarked that gold was heavier than feathers, otherwise how could it be said that a hook of gold was heavier than a cartload of feathers?'[22] (Immediately we have here another example that justifies the critique shared by Landes, Hsu, and myself.) After 15 centuries, some written records emerged that had something to do with the specific gravity or density of liquids. The topic was the test of brine density, and the favourite testing probe was lotus seeds. Late in the eleventh century Yao Kuan wrote:[23]

> Everyday I tested the brine with lotus seeds. The heavier ones were selected for use. If brine can float 3 or 4 such seeds (out of 10) it is considered strong brine. If it floats 5 (out of 10) it is the strongest. Those seeds which float perpendicularly are preferred. If only 2 seeds float perpendicularly, or 1 perpendicularly and 1 horizontally, the brine is considered thin and poor.

If all groups of 10 lotus seeds thus selected were identical statistical samplings (in terms of the specific gravity of seeds — a rough, rule of thumb hypothesis), the number of floats could show the specific gravity of the brine. But what had the specific gravity of brine to do with floating a lotus seed perpendicularly or horizontally? Indeed this whole device sounds more like a test of lotus seeds than of brine! There is definitely no science here, and if it is viewed as a technology, it is false technology. But Chinese stagnation in the physical science is featured more saliently by what Needham further tells us:[24] 'Brine-testing methods of this kind are often afterwards referred to, with variations, as in the mid +12th-century *Nen Gai Zhai Man Lu* of Wu Cheng. And they have continued in use down to our time.'

A recent Chinese paper explained two very terse passages — one of ten Han characters, another of 17 characters — in the Mohist Canon, and ascertained that they were about buoyancy and 'conform to Archimedes' principle of floating bodies which was discovered later'.[25] One passage says: 'An object of big body floats high on water. It is explained by equilibrium.' Another says: 'There is an equilibrium between the body and its submerged part. It is not the flatness of the body that decides its shallow submerge. It is as if there is a five to one exchange.' We may believe that 'a five to one exchange' is a better formulation than the whole story of weighing an elephant. (Although the former is generally incorrect whereas the latter is physically correct.) Indeed, both Mohist statements are couched in general terms. But they are too brief, and lack exactness.[26] One macro-historical fact is clear: Mohist discourse on buoyancy was not succeeded by any investigation until much later, and not as a subject of working science, but as a subject of the history of science. Many Chinese ideas and inventions share this fate: that is what we mean by Chinese scientific stagnation.

In order to reach a general principle about weight reduction of an immersed body, Cao Chong's episode could serve as a propitious starting point. But several more steps — by no means easy ones — are needed. First, it should be decided that the weight of the water displaced by the elephant is equal (not just proportional) to the weight of the elephant. Then the procedure of weighing an elephant with a floating boat should be formulated in abstract terms: the total weight of a floating boat and the things in it are balanced by replacing an amount of water of equal weight. But a floating body is not an immersed body. Therefore, empirical knowledge on the latter is needed; according to Needham, the Chinese were familiar with it — a situation, as we have seen, that depends on an inconclusive, indeed very dubious, explanation of what some Han technicians were doing. Next, a general formulation to summarise the empirical knowledge is needed, a formulation which is now accepted as the Archimedes' principle.

Finally, the principle should be explained theoretically, that is in terms of hydrostatic pressure, a step which was accomplished before the middle of the seventeenth century in Western Europe. In other words, if Chinese hydrostatic development were to contribute to the 'mechanical revolution' — that is to say, to catch up with Western development up to the eve of the Newtonian culmination — it required: (1) a general formulation about a floating body; (2) a general formulation about an immersed body; and (3) an explanation that unifies these and other hydrostatic facts (in particular those summarised in the Pascal principle). It is now clear that even with the most favourable interpretation of the practice of Han technicians, the actual Chinese situation in this respect can only be described as empiricist and underdeveloped, and remained so for a long time.

While reviewing Needham's work, I found that my idea of Chinese inertia resonated unexpectedly with a notion that emerged in a different context. D. Joravksky criticised P. Feyerabend for praising the Maoist sponsorship of traditional Chinese medicine while keeping silent about its obvious follies. In order to illustrate some 'critical questions' Joravsky used the case of 'tadpole contraceptive':[27]

> That rude defeat of belief by collision with physical reality raises critical questions . . . Was there some dreamy quality in the traditional Chinese mentality that permitted a persistent belief

> in a contraceptive method which could so easily be proved ineffective? And is there some harsher quality in the mentality of scientific physicians, some dream-destroying method, which persistently brings them closer than traditional rivals to objective tests of truth?

Have not we seen clearly this 'dreamy quality in the traditional Chinese mentality' in the recognition of facts and principles about buoyancy? The situations are not exactly the same, because in the mechanical context some ideas were correct (such as Cao's weighing the elephant), some partly incorrect (such as testing the specific gravity of brine with lotus seeds). But, in the case of correct knowledge, no one bothered to generalise it; in the case of incorrect understanding and inaccurate practice no one tested, criticised and improved it. Anyone who has had direct contact with unlettered or little-educated Chinese peasantry, (yes, present-day rural folk), knows that their major 'intellectual' interest is to carry on with some old hearsays rather than to carry out experiment. This, I believe, is a universal mentality, which might be labelled 'traditionalist'. When Galileo carried two different weights to mount the Leaning Tower at Pisa, he behaved as a revolutionary. Somewhat later, in China, as N. Sivin has told us, when Wang Xi-chan, Mei Wen-ding, and Xue Feng-zo studied Jesuit planetary astronomy with a critical and creative attitude, they also behaved as revolutionaries; although that does not mean a 'scientific revolution' occurred then in China. Joravsky was correct to indicate the 'harshness' of modern science, or the 'dream-destroying' quality of objective tests. When the discovery of Z^0 particle was announced in Geneva in June 1983 (in addition to the confirmation of W particles less than a year ago), two of the four 'basic' physical interactions — electromagnetic and weak — were officially amalgamated. For physical non-reductionists this is indeed such a harsh, 'dream-destroying' fact.

In terms of intellectual competence, learned Chinese, like learned people the world over, were certainly capable of using critical reasoning and could be convinced of objective tests as epistemological criteria. Contemporary with the astronomers Wang, Mei, and Xue, in China, schools that sought textual and archaeological evidence in historiographic research flourished. None the less, these schools could not compete with, still less replace, the traditional orthodoxy, which basically demanded unconditional acceptance among its followers. In the long run, traditional, or rural, mentality had a strong influence on the average Chinese attitude, because at any given time in traditional China, the ratio of the peasantry to other categories of people was disproportionately large. One can imagine that a man who advocated an experimental, critical methodology would look an eccentric to the public, as well as to his peers.

5 The Mohist 'Principle' of The Lever

The Mohist formulation of the principle of the lever is scientifically better than its buoyancy counterpart. It requires no basic corrections:[28]

> As for the steelyard, let a quantity of material and a weight be balanced, the distance between the fulcrum and the point where the material is suspended being shorter than the distance between the fulcrum and the point where the weight is suspended. This will then be the longer. If now to both sides the same weight is added, the weight must go down.

Three centuries later, in Liu An's *Huan Nan Zi* the following passage occurs:[29]

> Therefore if one has the benefit of 'position', a very small grasp can support a very large thing. That which is small but essential can control that which is wide and broad. So a beam only 10 *wei* long can support a house 1000 *jun* in weight; a hinge only 5 inches in length can control the opening and closing a large gate. It does not matter whether the material is large or small. What matters is its exact position.

Such a passage can hardly be thought of as a sign that indicated a wide understanding of 'the whole theory of equilibria as stated by Archimedes', as Needham contends. It was obviously a retrogression in comparison with the Mohist text, which strove to be general and quantitative. By the wording of the *Huai Nan Zi* passage we even do not know the relative configuration of the beam and the house, or of the '5 inch hinge' and the 'large gate'.

We have already remarked (see note 26) about one of the

linguistic difficulties of the archaic Chinese written language for scientific discussions, namely its extreme brevity. The language of Liu An's time was different from that of the Mohists which was less ornamental, but still no better from the point of view of scientific expression. I think, on the whole, that the written language and style of the long period of traditional China does not suit scientific or technological purposes because it was excessively concise; it deliberately omitted punctuations (also, Han characters do not have capitals); and it overemphasised aesthetic symmetry, poetic effects, and classical allusions. When this language describes natural phenomena or engineering feats, more often than not it sounds like either an exaggerating poet or a superstitious country cousin. Every classical Western language underwent some kind of modifications as it was adapted to communicate exact and experimental sciences. One well-known instance was the advocacy of a clear, straightforward style by the newly established Royal Society of London. Conversely, we may say that the long-lasting symbiosis of Chinese linguistic archaism and scientific inertia forms the other side of the Eurasian coin.

Owing to its emphasis on aesthetic effects, archaic Chinese had a large baggage of literary idioms. For instance, we have hundreds of the rhythmic 'four character phrases'. They are rhetorically effective; but a developing science needs a forthright and flexible language that experiments constantly with new things and new ideas. The use of archaic idiom, in addition to stereotyped syntactic and stylistic rules, poses a formidable barrier to expressing objects, processes, spatial configurations, and causal relations specifically, concretely, quantitatively, and rigorously. Moreover, Needham's illuminating discussion on the mutual influence between legal laws and natural laws ('natural laws' and 'laws of Nature' in his terminology) sheds some light on the topic of language-science relationship. Since the Chinese traditionally had a 'great distaste for precisely formulated abstracted codified law',[30] Chinese linguistic archaism remained unaffected either by scientific rigour and technological detail, or by jurisprudential deliberation and disputation.

6 Chinese Geometrical Optics

After Mo Ti, Chinese geometrical optics vanished for more than a millennium until Shen Kua (1031–95) studied it, unaware of any conclusions made by the Mohists. In the middle of the fourteenth century, Zhao You-qing carried out some brilliant experiments to prove the rectilinear propagation of light and to explain the image formed through a small hole,[31] again unaware of his predecessors, either the Mohists or Shen Kua. In the realm of physical inquiry, the lack of professionalism and the lack of professional inheritance are evident. But we may also raise a related question: does the lack of professional inheritance imply a lack of motivation to make deeper inquiries? It was because traditional China did not provide the necessary intellectual zeal to sustain enduring systematic, rational, and causal inquiries about nature, that we encounter repeated instances where some topic was picked up by an interested scholar but then was subsequently disregarded. That is why Chinese geometrical optics, and physics in general, could never proceed beyond several elementary and imprecise statements. On the other hand, mathematics in traditional China posed, to a certain degree, a contrast to physics.

7 Mathematics — Traditional China's Pride

In traditional China there had been a persistent political incentive for pursuing mathematics, which might be viewed as a calendrical auxiliary. The management of a vast agrarian society, together with the frequent changing of reigns and dynasties, put a premium on calendrical regulations. Sometimes the social commotion caused by a peasant rebellion or natural catastrophe would require a new calendrical beginning. In 2,000 years, traditional China had more than 40 calendrical reforms and another 50 unsuccessful proposals![32] Calendrical reforms required the service of mathematicians:[33]

> Turning to social factors, it is striking that throughout Chinese history the main importance of mathematics was in relation to the calendar. It would be hard to find a mathematician in the *Zhou Ren Zhuan* who was not called upon to remodel the calendar of his time, or to help in such work.

(*Zhou Ren Zhuan* is *Biographies of (Chinese) Mathematicians and Astronomers*, written by Ruan Yuan, printed in 1799). During the thousand-year period between the Han and the Tang there developed the renowned 'Ten Mathematical Manuals', which were collected and issued in the Song. In the Department of Mathematics (Suan Shu Ke) of the State University (Guo Zi Jian), these ten manuals were used as textbooks. The earliest of the ten manuals was *Zhou Bei Suan Jing*, a work which originated and stabilised between the sixth and first centuries BC. Although in this book the 'theorem of Pythagoras', and the diagram used, were expressed in an equivalent of

$$3^2 + 4^2 = 5^2$$

it is clear that the method of proof can be easily generalised to vindicate the 'Pythagorean theorem' generally — as Needham declares, 'The Chinese proof of the Pythagoras theorem was indeed a proof.'[34] Curiously enough, while no Chinese mathematician had ever made the step towards the generalisation that

$$a^2 + b^2 = c^2$$

holds true for *any* right-angled triangle, the truth of this theorem has been taken for granted, and was put to great use, most notably in Liu Hui's brilliant evaluation of pi.

In the third century (AD) Liu Hui made comments on *Jiu Zhang Suan Shu*, another book of the ten manual series. On the ratio of the circumference of a circle to its diameter, he wrote that the ancient belief about the circle that 'perimeter 3 diameter 1' is inaccurate; that 3 to 1 is in fact the ratio between the perimeter of the largest hexagon inside a circle to the circle's diameter. In order to calculate pi, Liu Hui pointed out that one can obtain the perimeter of a 2n-side-regular-polygon from the perimeter of an n-side-regular-polygon (both polygons are the largest inside the circle). The 'Chinese Pythagorean theorem' served as his theoretical basis, which can be sketched with the aid of a diagram:

In the diagram a is one segment of an n-side-regular-polygon, and b, one segment of a 2n-side-regular-polygon. If a is known, then the Pythagorean difference of the radius r and a/2 is known. By the 'Chinese Pythagorean theorem', we can obtain b by

$$b^2 = \left(\frac{a}{2}\right)^2 + \left(r - \sqrt{r^2 - \left(\frac{a}{2}\right)^2}\right)^2 \circ$$

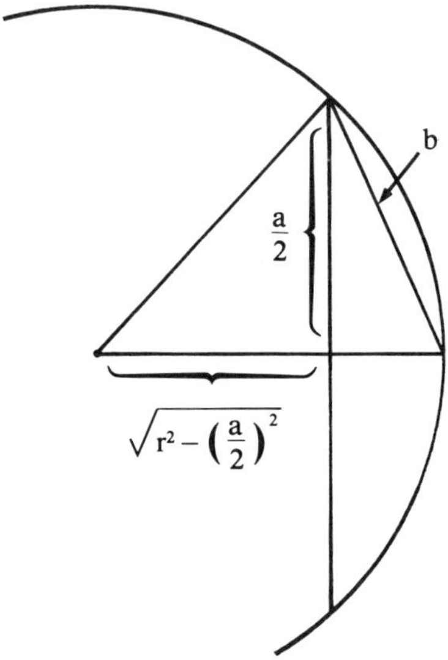

Liu Hui carried on his process of calculation up to a regular polygon of 3,072 sides and had as his best value of pi, 3.1416.[35]

Two centuries after Liu Hui, in the fifth century, also in the period when the Confucian grip over China was relaxed, Zu Chong-zhi and his son, Zu Geng-zhi, further pushed the evaluation of pi to a new order of accuracy. According to them, an 'excess value' of pi is 3.1415927, and a 'deficit value', 3.1415926. This was a result not equally anywhere until the end of the sixteenth century, a remarkable Chinese scientific champion! Their method was lost; but historians of mathematics believe that they used Liu Hui's procedure, with which they should have used 'counting-rods' to do an enormous number of arithmetic operations for a 12,288-side-polygon and a 24,576-side-polygon, indeed a superhuman effort. Needham correctly points out that at the time of Liu Hui, 'the Chinese had more than caught up with the Greeks' — Archimedes obtained his pi by working with a 96-sided polygon; then with the Zu father and son, the Chinese had again moved 'ahead for a thousand years'.[36]

Some dramatic facts concerning the Zu family[37] illustrate two interesting points. First, there was no lack of professional inheritance in applied sciences such as mathematics and astronomy. (This was certainly also true of medicine and alchemy.) Zu Chong-zhi's grandfather, Zu Chang, was the Palace Architect (Da Jiang Qing) of the (Liu) Song court. Zu Chong-zhi himself was sponsored by a (Lui) Song emperor as an astronomer and engineer. He determined the solar year to be 365.24281481 days, which differs by only 50 seconds from the modern determination. His lunar cycle was 27.21222 days, or less than 1 second off modern calculation. But the narrations about his mechanical feats sound more like the *perpetuum mobile*. A second point is that political instability interrupted scientific traditions. Besides the calculation of pi, Geng-zhi, Chong-zhi's son, invented a method to calculate spheric volume; he constructed instruments which told time accurately by either measuring the shadow of the sunshine or the accumulation of dripping water. His son, Zu Hao, that is Chong-zhi's grandson, was also a talented mathematician. But he was killed in a political upheaval — the Hou Jing Rebellion. This tragedy ended a brilliant five-generation mathematico-technological family tradition.

After the 'counting-rods', the Chinese invented and developed the abacus. Chinese historians generally differ from Needham on the dating of the invention of the abacus.[38] The Chinese dating is much later than that of Needham (hence indicating a much lesser inertia on the part of Chinese computing science). Needham believes that a description dating from about AD 570 gives evidence of the existence of the abacus. Recent Chinese research on this problem asserts that as early as in 1279 a number of literary works show that the abacus was already popular (not as late as the end of the fifteenth-century, as Needham in 1959 believed the Chinese dating to be).[39] I think Mei Rong-zhao's argument about the absence of a full-fledged abacus in the sixth-century is convincing.[40] He points out that not only does the three-section structure differ from the later two-section structure, but more importantly, that there had not been the mnemonic rhyme that enables one to do multiplication and division with an abacus. It was, in fact, a different device. This is an important point, because it reminds us of a modern analogy in the very area of computing science: the crucial role of the software such as a mnemonic algorithm.

Some of the algebraic problems to which Chinese mathematicians contributed remarkably had their origin in calendrical studies, such as the 'indeterminate problem' in *Sun Zi Suan Jing*,[41] a fourth-century work and one of the ten manuals. According to Li and Yuan's analysis,[42] the question has its origin in determining the initial time of a calendar. The formulation of the mathematical problem was on the one hand put in various popular forms, and on the other hand, later in the thirteenth century, was generalised by the great Chinese mathematician Qin Jiu-shao into 'the Chinese theorem of remainder'. He systematically enunciated the general scheme of linear indeterminate analysis. Qing finished his *Shu Shu Jiu Zhang (Mathematical Treatise in Nine Sections*) in 1247, during the last stage of the Southern Song. His achievements suggest that there were intellectual interests in theorising exact sciences at that time. Guo Shu-chun thinks that it was mainly because of the inhibition posed by Neo-Confuciansism that after the Song and the Yuan mathematics declined.[43] It must have been a rapid decline, since when 'the Jesuits entered upon the scene (the sixteenth century), there was no one even able to tell them of China's past mathematical glories'.[44] Why so rapid a decline? As we have said before, mathematics in traditional China was basically an official enterprise. There were not enough autonomous roots of mathematical scholarship among learned circles. Correspondingly, the best achievements in the field appear to have been isolated — both spatially and temporally — successes of mathematical wizards. Therefore, once official encouragement declined, mathematics declined rapidly. An indirect support of this explanation is that these limited, though brilliant, mathematical feats could not bring a perceptible positive influence, the characteristic clarity and exactness of mathematics, to general Chinese scholarship or to the Chinese written language.

8 Mathematisation or Axiomatisation?

In concluding his section on mathematics, Needham discusses the general topic of 'Mathematics and Science in China and the West'. He stresses the emergence of modern science as a result of the mathematisation of natural study, in which campaign Galileo occupied a central position. Galileo is contrasted with a long list of practitioners of the fifteenth and sixteenth centuries, among whom Leonardo da Vinci was the most outstanding.[45] Needham thinks that Pledge hits the target when he says that in spite of all Leonardo's deep insight into nature and brilliance in experimentation, he achieved no further development because he lacked mathematics. Yet four pages later Needham said that Strong[46]

> had convincingly shown that before Galileo, and during his lifetime, mathematics was increasingly utilized by the practical technicians and artisans of whom we have already spoken. Some of these, such as Nicolo Tartaglia and Simon Stevin, were among the best methematicians of their time.

Needham insists that a crucial step of scientific progress had been made by Galileo, and not by any of the previous practitioners. Why Galileo? Because of his mathematics? (In fact, compared with Kepler, Descartes, and Newton, Galileo was obviously the least mathematical.) But his predecessors also had mathematics (Leonardo, Tartaglia, Stevin, and Galileo were born respectively in 1452, 1500?, 1548, and 1564). As a result, we readers are left in a logical quandary. In order to resolve this quandary, I think we ought to be aware that mathematisation could not provide the true 'magic touch',[47] but 'axiomatisation' did. Galileo and Tartaglia both did mathematisation, but one did it in the realm of basic physical science, the other in the realm of technology. The operations of looking for measurables and hypothesising their relations might be executed either on the level of basic phenomena, or on a derivative level. Science works on basic phenomena and arrives at conclusions which are universally valid but which may be remote from practical applications. That was the type of work Galileo and Kepler accomplished. By way of axiomatisation, methodologically like Euclidean geometry, science tries to embrace as many facts as possible so that all are ordered in a unified explanatory system. Galileo and Kepler did their respective axiomatisations; Newton axiomatised both of them.

Since basic physical relations are quantitative, a successful axiomatisation necessarily implies mathematisation. Galileo was very aware of this. Thus he wrote that 'philosophy' is written in mathematical terms and symbols. Descartes was also very aware of this when he said he would take 'God' and 'the mathematical order of nature' as synonyms. Mathematised disciplines impress people with their accuracy and exactness. Yet, an advanced axiomatisation possesses even greater intellectual appeal. Newton's synthesis, of course, was well equipped with both exactness and comprehensiveness, which, in turn, impressed eighteenth-century philosophers, and thus contributed conspicuously to the general progress of thought. Science did not directly contribute to engineering technologies until about a century and a half after the publication of the

Principia. But indirectly, the contribution of the Scientific Revolution — owing to a positive attitude towards intellectualism and rationality — could never be dismissed for a single instant.

The growth of modern science as a whole does not exclusively mean mathematisation. Up to the present, science as a whole is far from being totally mathematised, and will perhaps never be so. William Harvey's discovery of the circulation of blood formed a part of the Scientific Revolution of Galileo's time, yet it was not written in the language of mathematics. But the human intellect always tacitly accepts some fundamental assumption: explanation means accounting for a wide range of phenomena with a limited number of elements. The Yin and Yang, or the Five Elements, tried to work this way. They were the earliest instances of attempting axiomatisation in China. Nevertheless, in the subsequent long centuries of Chinese history, we can seldom find the dimmest consciousness of scientific axiomatisation. (In making this statement, the first specific case in my mind is again the non-acceptance of Euclidean geometry. See note 16.)

We can further drive home the difference between axiomatisation and mathematisation by studying another passage from the same section of Needham's work. He agrees with Koyrè that before the appearance of Newtonian-Cartesian science, 'mathematics itself had to be transformed'. Then he lists the following Eureopean mathematical inventions that demonstrate the transformation:[48]

> In +1550 European mathematics had been hardly more advanced than the Arabic inheritance of Indian and Chinese discoveries. But there followed an astounding range of things basically new — the elaboration of a satisfactory algebraic notation at last by Vieta (1580) and Recorde (1557), the full appreciation of what decimals were capable of by Stevin (1585), the invention of logarithms by Napier (1614) and the slide-rule by Gunter (1620), the establishment of coordinate and analytic geometry by Descartes (1637), the first adding machine (Pascal, 1642), and achievement of the infinitesimal calculus by Newton (1665) and Leibniz (1684).

Yet as we know, Newton did not use the infinitesimal calculus to demonstrate his *Principia*.[49] His small bundle of axioms for mechanics comprised only his three laws of mechanics (the first, and an important special case of the second, had been discovered by Galileo), and the law of gravity. In order to account for all the theorems on motion with which Kepler and Galileo had provided him, none of the above-mentioned newly-invented mathematical tools, as specific aids, was necessary for Newton. But Euclidean geometry and Cartesian synthesis inspired him as methodological models. In developing exact science, axiomatisation — in general, the search for rational explanations — has been the core; mathematisation has acted as an indispensable tool.

9 Lack of Intellectual Activation and Creativity

In retrospect, it becomes clear that mechanics was 'modernised' in three steps: Archimedes' theorems of the lever and buoyancy; Galileo's and Kepler's theorems on a variety of mechanical phenomena; and Newton's synthesis. In the light of this scheme, ancient Chinese mechanics, as we have discussed before, had fulfilled the first half step, but never moved beyond that independently. I agree with Mikami and Fu Si-nian[50] that the great deficiency in old Chinese mathematical thought was the absence of rigorous proof, in particular the absence of a system of deductive geometry. This configuration correlates with the lack of formal logic and the dominance of associative (organicist) thought. From our Sino-European comparison, it is clear that the deficiency was not just in mathematics; it hindered the development of modern science as a whole. In other fields of science, the Chinese way of thinking generally lacked accuracy in defining, exactness in formulating, rigour in proving, and logic in explaining. The *Ben Cao Gang Mu*, the great pharmacopoeia of the sixteenth century, records the 'tradition' of preparing a toxin this way: placing many toxic insects in a closed vessel and allowing them to remain there until one had eaten all the rest — the toxin was then extracted from the survivor![51] But this 'method' presupposes a number of assumptions, including that insects are interested in fighting and eating each other; that they form a predatory hierarchy in which the last survivor 'concentrates', or 'collects', rather than 'detoxinates', all the poisons present in the other insects. Who had proved these hypotheses and under what control? There is no mention. Passages like this are certainly not science, but quack literature. On the other hand, it might serve as a working hypothesis, ready to be improved and upgraded. However, owing to the absence of an academic atmosphere activated by criticism and counter-criticism, and the presence of the 'dreamy mentality' that permitted the persistence of disprovable hearsay (see Section 4 above), no one would treat it thus.

Needham made the following suggestion to explain China's lack of mechanical development:[52]

> The Chinese should have been interested in mechanics for ships, in hydrostatics for their vast canal system (like the Dutch), in ballistics for guns (after all, they had possessed gunpowder four centuries before Europe), and in pumps for mines. If they were not, could not the answer be sought in the fact that little or no private profit was to be gained from any of these things in Chinese society, dominated by its imperial bureaucracy?

My explanation is summarised in the so-called principle of degree of intellectual activation and creativity. What Needham mentions above was undoubtedly important. On a derivative, or phenomenological, level, it was because Chinese society was far from being fully activated intellectually — especially on matters that finally turned out to be exact sciences and technological engineering in a modern world. Owing to the lack of competition in creative thinking, criticism on the basis of objectivity, tests in terms of empiricism, and checks by logical reasoning, traditional Chinese science generally stayed at an archaic or underdeveloped status. A distant and vague analogy would be accepted as a piece of convincing evidence. In so far as explanatory scholarship is concerned, traditional Chinese science could boast little. There was indeed plenty of metaphysics, but very little physics.

10 Chinese Astronomy Stayed Empiricist

Needham describes astronomy in traditional China as an orthodox, official, and Confucian science,[53] which was characterised by its early recording of sun-spots, 'new-stars', comets, and meteors. Like calendrical revisions, these activities were part and parcel of bureaucratic necessity. As I have mentioned before, calendrical regulation was required to govern a vast agrarian state. On the other hand, persistent interest in recording any celestial anomaly or eccentricity may reflect China's characteristic political instability. For 3,000 years, Chinese astronomy remained empirical and observational.[54] Earlier we mentioned that Aristotle's mistaken 'law of falling bodies' was scientific. Analogously, we may say that Ptolemaic planetary astronomy was scientific, in the sense that the solar system itself was correctly isolated and an orbital explanation was attempted — it was an early combination of mathematisation and axiomatisation. Thanks to this common scientific ground, Ptolemy's *Almagest* was replaced by Copernicus' *De revolutionibus*, just as Aristotle's 'law of falling bodies' was replaced by that of Galileo's — although the latter was a case of much smaller scale and subject to direct test. We have already outlined the development of mechanics, and examined Chinese mechanical underdevelopment. It seems a similar situation existed in Chinese planetary (or explanatory) astronomy. There was only a half step, or even less:[55]

> In spite of so much accurate observation, Chinese study of planetary motion remained purely non-representational in character. Unlike that of the Greeks, in which the geometry of circles and curves was so prominent, it perpetuated the algebraic treatment of the Babylonian astronomers such as Naburiannu and Kidinnu, and never sought a geometrical theory of planetary motions.

Concerning Chinese planetary astronomy, Sivin writes:[56]

> Prior to Shen's time little effort had gone into predicting the apparent motions of the planets, which lacked the immediacy of solar and lunar phenomena. This was, in fact, an omission that Shen seems to have been the first to confront.

Shen is Shen Kua (1031–95), a polymath whom we have met before in connection with his interest in geometrical optics. He once headed the Astronomical Bureau of the Northern Song Dynasty (960–1126). The plan with which he and his collaborator Wei wanted to confront planetary astronomy was 'exact coordinates read three times a night for five years'. But, owing to the 'antagonism within the Bureau', such a plan could not be carried out. Consequently, 'Shen and Wei had no recourse but to produce a conventional planetary theory based mainly on old observations.'[57]

For the major features of the subsequent development, let us quote Sivin again:[58]

> Very gradually from the Yuan period (1279–1368), it came to be little practised outside the Astronomical Bureau, which was dominated by foreign technicians. By 1600 no one was able fully to comprehend the old numerical equations of higher-order, prototrigonometric approximations, applications of the method of finite differences, and other sophisticated techniques.

Then came the age of Wang Xi-chan (1628–82), a hitherto comparatively obscure astronomer, but for recent Sinophile revisionists one of the leading figures of the 'Chinese scientific revolution' in the seventeenth century:[59]

> Wang, Mei Wen-ding, and Xue Feng-zo were the first scholars in China to respond to the exact science and to shape their influence on their successors. They were, in short, responsible for a scientific revolution. They radically reoriented the sense of how one goes about comprehending the celestial motions . . .

The general attainments of this 'scientific revolution' are reflected on the one hand by a conjectural remark by Xi Ze-zong:[60]

> We can imagine, if Wang Xi-chan had only come upon [Copernicus' *De revolutionibus*, Galileo's *Dialogue*, and Kepler's *Epitone astronomiae Copernicanae*, all of which the missionaries kept for their private use in Peking], how much greater his contribution to astronomy would have been.

And on the other hand by actual scientific performance in China in the past three centuries. We have discussed the scientific contents of the European Scientific Revolution. The seventeenth-century Chinese 'scientific revolution' certainly does not match this objective criterion. Neither can it be so qualified because of its revolutionary social consequence. People understand the Scientific Revolution as one of the greatest historical movements because it positively and permanently influenced the world. Has the so-called Chinese 'scientific revolution' wrought any comparable changes for China? Evidently not. If there had been an even dimly successful 'scientific revolution' in China 300 years ago, the general condition of the country, as well as the present scientific situation, would not cut so mediocre a figure. As a nation that comprises a considerable portion of the world's population and that has a long and remarkable culture, China should at least lead the world in several fields of scientific and scholarly studies. Alas, what China can definitely boast about today is her 'science' of playing ping-pong. Therefore, by the 'terminal condition', or the 'output', we deduce retrospectively that there had not been a 'scientific revolution'. We noticed that the system of unscientific, or even anti-scientific, civil examinations was abolished only as late as 1905. Since the turn of the century, students were sent abroad. But in the first few decades of this century the acceptance of science still had to struggle against strong reluctance. Modern science takes root in Chinese soil very slowly, even today. During my middle school years between 1947 and 1952, the great majority of scientific textbooks were translations of American books. We learned from history that in the middle ages, the European textbooks were originally Arabic. In many ways, for a century China has been in a 'translation movement', and in the past three decades, an authoritatively controlled 'translation movement'.

11 Theories on Tides; Meteorological Instruments; Cartography

Science advances by critical inheritance, that is through debates. The lack of debate was in fact a part of the lack of professional inheritance in scientific matters. More generally, as I have mentioned before, the situation reflects the lack of intellectual activation and creativity.[61] The whole book of Wang Chong's *Lun Heng* (AD 83) is imbued with polemics, which makes it unique in early Chinese science. He surpassed his predecessors in his understanding of the distinction between the aqueous circulation in the atmosphere and the vastly distant starry firmament,[62] of thunder and lightning,[63] and of tidal bores.[64] The causal understanding of sea tides progressed slowly: in the ninth century Lu Chao wrote an 'Essay on the Tides', in which he states that 'the sun produces tides, but the moon determines their dimensions'[65] — a fine and false balance of the Yin and Yang. More than three centuries passed before Shen Kua, in his now famous *Brush Talks from Dream Brook*, 'took Lu Chao to task for his theory that the sun was involved in th causation of tides'.[66]

The Chinese were early in the construction of meteorological instruments, such as hygrometers, using charcoal or feather,[67] and the weather cock. But Needham might have inordinately magnified the significance of the latter: it is 'perhaps the oldest of all pointer-reading devices, the importance of which in the philosophy of the natural sciences requires no emphasis'.[68] The majority of pointer-reading devices function according to Hooke's law of elasticity; a visible indicator, such as one's finger, is certainly not an important invention.[69]

The picture of the first thermometer in China,[70] a Jesuit importation, reminds one of the growth of modern science as a twofold process of the growth of the number of measurable quantities and the growth of measured accuracy. Measurements of other than an economic nature, in the sense of both new measurables and new accuracies, demand a true conceptual leap. The perception of the necessity, and the recognition of the possibility, of these measurements demands the extension of imagination to remote analogies and abstraction; the physical realisation of these measurements, such as the early thermometers, requires the construction of devices which are sensitive in several aspects. In Western Europe the measurement of small time intervals, temperature, small and distant space intervals (microscope and telescope), etc. were the great achievements of successive generations of Renaissance scholars who combined intellectuality with craftsmanship. In China, Su Song (1020–1101) of the Song Dynasty, Guo Shou-jing (thirteenth-century) of the Yuan Dynasty, Li Shi-zheng (1518–93) of the Ming Dynasty, etc. were men of this type. The first two were astronomer-engineers; the last, the renowned author of the *Ben Cao Gang Mu*, which we quoted before for its preparation of a 'virulent poison', contributed nothing to quantitative sciences.

Cartography, because of its susceptibility to repetitive tests and its openness to positive improvements, was no doubt the best part of traditional China's 'science of the earth', which in Needham's work comprises cartography, geology, seismology, and mineralogy. China had been very early in producing maps. The wars in the period of the Warring States served as an effective incentive for cartographic improvements. Intellectual motivation in this aspect was for once adequate. No one seriously questions the validity of the existence of pre-Qin, that is pre-221 BC, maps, yet until ten years ago no actual map was known which substantiated this early Chinese achievement. Needham's volume III was published in

1959; it could mention no extant pre-Qin maps. Then in 1973, a bumper harvest of archaeological excavation incidentally solved the question satisfactorily. Near Changsha, Hunan Province, an aristocratic tomb of 168 BC was opened. Among many valuable artefacts were three well-drawn maps on silk fabric.[71] Although they were buried half a century after the post-Qin era began, it is understandable that historians' cartographical curiosity was greatly satisfied by these precious relics.

Because of inherited high standards, personal practice, and hostility to low quality maps, Pei Xiu could expound his six cartographical principles in AD 267. After Pei Xiu, an important stage occurred during the Tang Dynasty — as for many other things — which 'saw a great development of map-making'. 'The wide extensions of the empire which were added early in the dynasty stimulated the mapping of Central Asia.'[72] Apparently progress was not only made in quantity, but also in quality. The survey instruments, shown on page 570 of Needham's book, look so modern; yet according to Needham they were much earlier (759, in the middle of the Tang) than their publication (1044).[73] Then the Song Dynasty left us some real masterpieces, 'the most remarkable cartographic work of its age in any culture'.[74] It makes sense when we notice that the first masterpiece was produced almost contemporaneously with the publication of the above-mentioned drawing of survey instruments. The Ming Dynasty saw the achievements of Zhu Si-ben (1273–1337), a man of respectable scientific integrity.[75] But when the Jesuits arrived in the sixteenth century, it was obvious that Western cartography was in advance of that of the Chinese. Needham did not analyse the reasons, which by an educated guess should be the inadequacy of indigenous Chinese geometry and mathematics (see note 16), together with the lack of knowledge that the earth was a globe. At any rate, it was Matteo Ricci who came from Western Europe to China and then made his famous world map of 1602.[76] In other words, it was not a Confucian scholar who left China for Europe, although in Chinese history there had been Buddhist pilgrims. The eunuch admiral of the Ming, Zheng He, was an official ambassador as well as a Confucian missionary. From comparisons such as these we realise that a Confucian missionary could only be delegated by a government, because Confucianism is essentially political. It should be useful for us to recall this contrast when we are considering China's non-development of science. In comparison with politico-ethical theorisation (let alone political pragmatism), religious zeal is spiritually much nearer to a scientific mentality.

12 Some Questions About Chinese Seismology

In the field of earth science, I think the alleged Zhang Heng seismograph is debatable on several accounts. I certainly do not justify my scepticism with the same arguments as Zhou Mi (thirteenth century) or He Xiu (eighteenth century).[77] Neither can I argue experimentally. But I will try to do it macro-historically. First, as described in the *Hou Han Shu* — probably our sole historiographic evidence — some recordings were impossible: a single instrument cannot tell *where* the earthquakes come from. Secondly, some other recordings claim to show that Zhang Heng's mechanical seismograph was more sensitive than present-day electronic instruments. Thirdly, Zhang Heng's success — if true — shows that his engineering design and Han craftsmanship cannot be matched by some modern scholars' collective efforts. Fourthly, Zhang Heng's seismograph, as a physical object, disappeared, apparently shortly after its creation. It seems that the third point deserves more discussion. A 1980 paper by a staff member of the Institute of the History of Natural Sciences in Beijing includes a photo of the reconstructed Zhang Heng seismograph that shows its aesthetic appearance, but says nothing at all about its practicality as an instrument. From Needham's book we learn that in the Seismological Observatory of Tokyo University, Imamura and Hagiwara obtained some positive results: some balls of the instrument dropped, anyhow. Logically, the next step of research should solve questions such as these: Given the materials and processing technique available to Zhang Heng, what level of accuracy could his seismograph attain? Was it possible for it to single out one particular direction?

For the theme of this essay, the moral of the Zhang Heng seismograph is that regardless of the degree of its success or its magical effects in a superstitious dynastic court, neither its physical existence nor its description ever inspired a later Chinese scientist to imitate and reconstruct it, still less to improve it. The description inspired only critiques, as noted by Needham. Zhang Heng's seismographs is another good illustration of one of our previous characterisations of Chinese scientific stagnation: it had rarely succeeded as part of working science, but is now discussed as part of the history of science. But specific reasons for scientific discontinuation might differ. In the case of the Zhang Heng seismograph, I believe it was because it could not measure up to its scientific claims.

With regard to ancient explanations of earthquakes, Needham's section on seismology contains a revealing comparison between an ancient Chinese theory and ancient Greek theories. The Chinese theory, which is described in the *Shi Ji (Records of a Historian)* (ending in 99 BC), says:[78]

> The dynasty of the Zhou is going to perish . . . When the Yang is hidden and cannot come forth, or when the Yin bars its way and it cannot rise up, then there is what we call an earthquake. Now we see that the three rivers have dried up by this shaking; it is because the Yang has lost its place and the Yin has overburdened it. When the Yang has lost its rank and finds itself (subordinate to) the Yin, the springs become closed, and when this has happened the kingdom must be lost . . .

After this quote from the *Shi Ji*, Needham describes the continuous development of the theory of 'the imprisoned Yang' until the Song Dynasty, as late as the end of the thirteenth century — and it might have persisted even longer. About ancient Greek theories Needham uses Lones's summary:[79]

> Anaxagoras believed that earthquakes were caused by excess of water from the upper regions bursting into the under parts and hollows of the earth; Democritus thought that this happened when the earth was already saturated with water, and Anaximenes suggested that the shocks were caused by masses of earth falling in cavernous places during the processes of drying. Aristotle himself in the -4th century attributed the instability to the vapour generated by the drying action of the sun on the moist earth, and to difficulties met with by the vapour in escaping.

Although Needham reminds his readers that Chinese theories were no more primitive than those which the ancient Mediterranean world entertained on the subject of earthquakes, we could hardly help noticing two striking differences between them. One is the strong political association in the Chinese natural theory.

Consequently it acquires an authoritativeness that intends to compel people to accept their political conclusions. (Is this one of the keys to understand the characteristic *non sequitur* in Chinese reasoning?) Another problem about the Chinese theory is the undefined 'Yin and Yang'. Looking at the terse summary of Greek theories, one admires by contrast their unadulterated materialism: 'excess of water', 'saturated with water', 'masses of earth', or 'the vapour generated by the drying action of the sun'.

I am aware of the long survival of such erroneous concepts as 'phlogiston' and 'ether' in Western scientific development. But they differ saliently from the Yin and Yang, because their physical and chemical functions and characteristics were well hypothesised. In the eighteenth century the 'phlogiston', and as late as the beginning of this century the 'ether', both played their parts in hectic times of experimental and argumentative sciences. When science pushed them out of its realm, as if by Newtonian reaction, they pushed science ahead. In China, the Yin and Yang appeared early, and were backed by political as well as intellectual authority. As if mired in a vicious circle, Chinese science could not expel them, hence could not make the necessary breakthrough.

13 A Scientific Musical Scale — Another Chinese Feat

As we have mentioned before, some aspects of acoustics provided the foremost successes of traditional Chinese science. Not accidentally, Dong Zhong-shu's description on resonance reads like experimental physics. In the following centuries, the study of the musical scale in particular, both experimentally and theoretically, developed steadily. Two sections in Needham's book, 'the search for accuracy in tuning' and 'the recognition of sound as vibration' especially impress the reader with the sequences of experimental developments. Yet the quintessence of Chinese quantitative science was the theoretical recognition of the scale. Since we limit ourselves to traditional China, we will not follow Needham and discuss the question 'Pythagoras or Lin Lun'. Neither are we interested in the question of whether both sides had acquired their inspiration from the Babylonians. The relevant fact for us is that traditional China had inherited a numerical procedure to construct a scale in an approximate octave. This procedure Needham calls 'the Chinese spiral of fifths';[80] in ancient Chinese work it is named 'plus and minus to three divisions',[81] apparently an ancient name since the Spring and Autumn (722-480 BC). By this procedure one constructs a fairly good twelve-note scale by multiplying the basic (frequency) by the fractions 2/3 and 4/3. As a pentatonic scale, the Chinese gamut conforms to the Pythagorean scale exactly, but cannot make an exact octave: the discrepancy is that between 1/2 and 262144/531441. Chinese musical practice had never developed the sophistication to use all twelve notes; yet this numerical procedure indicates that there were early Chinese scholars who were theoretically-minded, and who wanted to generalise and mathematise empirical facts. More than that, the full significance of this work is shown by the persistent effort over many centuries to improve it. As a result we have a lengthy list of musicologists who belonged to the Han, the (Liu) Song, the Liang, the Jin, and the Song Dynasties. Their effort aimed at making an equal-tempered scale, the essence of which is that the ratio of one note to another depends only on the difference of their numbering (and an exact octave would ensue automatically). The 'plus and minus to three divisions' is not equal-tempered. Even if we divide an octave into twelve equal divisions (by 'length'), these divisions do not form an equal-tempered scale, let us, for instance, examine the ratio between the eleventh and ninth notes and that between the ninth and the seventh:

$$(11/12)/(9/12) = (11/9) = 1.22\ldots \neq (9/12)/(7/12) = (9/7) = 1.28\ldots$$

The persistent effort was finally crowned in the hands of Zhu Zaiyu, a Ming prince.[82] He solved the question by pointing out that if the twelve divisions between for instance 1 and 2 (convenient to be expressed in the following series) is made at these numbers:

$$2^{\frac{0}{12}}(=1),\ 2^{\frac{1}{12}},\ 2^{\frac{2}{12}}\ \ldots\ 2^{\frac{11}{12}},\ 2^{\frac{12}{12}}(=2)$$

then the ratio of any two notes, for example the ratio between numbers m and n, is

$$2^{\frac{m}{12}}/2^{\frac{n}{12}} = 2^{(m-n)/12}$$

which depends on the difference (m−n) only, and has nothing to do with the positions of m and/or n themselves. In this way the scale is equally tempered, and an instrument of this scale can play any key by an adjustment of 'parallel displacement'.

Because of this 'crowning achievement of China's two millennia of acoustic experiment or research', Needham acclaims Zhu Zai-yu as 'one of China's most distinguished mathematical and musicological scholars'. Zhu's work was printed in 1584, much earlier than the first European mention of equal temperament (1636).[83] In fact, Needham suspects strongly that European equal temperament was the result of learning from Zhu Zai-yu (see also above, Section 4).[84] Zhu's work was also a few decades earlier than the works of Kepler and Galileo. By its general character, equal temperament reminds one of Galileo's discovery of the mathematical formula of the simple pendulum. If we recall our comparison between Galileo and Tartaglia, we would agree to group Zhu and Tartaglia together. Zhu did not try to mathematise basic mechanical laws. He could not lead a true scientific revolution in China. Traditional China lacked the general conditions for the emergence of a mathematised mechanics. Obviously, one of the necessary conditions for the Galileo–Kepler–Newton culmination was the availability of positive results from a wide range of mechanical studies. Zhu was historically too unique a talent, and his scholarship produced too limited a masterpiece. The personal background of Zhu reminds us of another Chinese phenomenon: throughout traditional China a number of scientists were imperial descendants, or men living in an age of dynastic transition, or other types of men who were not obliged to participate in civil examinations.

14 Chinese Magnetism versus Gilbertian Magnetism

Finally we come to Chinese magnetism to finish our review of the physical sciences in traditional China.

A scientific summary of pre-modern magnetism can be fairly compact. As phenomenological knowledge, it is basically composed of two parts: magnetic attraction (and repulsion), and magnetic directivity. Both the Chinese and the Greeks early recognised some facts about the first part, but the discovery of the important second part was an early Chinese glory. Besides the south-north pointing property of the magnetic needle, the more sophisticated facts in the category of magnetic directivity are: first, magnetic declination, that is, the magnetic needle does not point exactly to the geographic south; second, local variations due to the presence of rocks of magnetic nature; third, the inclination, or magnetic 'dip' — a magnetic needle does not point exactly in a horizontal plane; and fourth, that the earth itself is a big lodestone — a fact so remote from direct observation that it may also be categorised as a theory.

The discovery of the magnetic declination was also a Chinese first. Shen Kua wrote in his *Brush Talks from Dream Brook*:[85] 'Magicians rub the point of a needle with the lodestone; then it is able to point to the south. But it always inclines slightly to the east, and does not point directly at the south.' This observation antedates by 400 years the traditional time of the discovery of declination in the West, by Columbus in 1492. More than that, Needham could trace it back at least to the middle of the ninth century and ascertain the order of the described declination.[86]

As for local variations, Needham believes the following 1436 Chinese passage shows the first awareness of them:[87]

> There is a seamen's saying 'To the north we are afraid of the Seven Island: to the south we fear the Kun-lun (Pulo Condor Island).' At these places the needle may err, and if that happens, or the steering is inaccurate, both men and ships will be lost.

Traditional China failed to attain the knowledge of the next two facts, which demand either experience in oceangoing navigations, or careful measurements, or theoretical imagination. Needham states that[88]

> So far as it has been possible to ascertain, the vertical component of the earth's magnetic field (magnetic dip or inclination) was never discovered in China. According to Mitchell the first observation of it was due to Hartmann in 1544, and the first correct estimate of it to Robert Norman in 1581.

So, this course of development chronologically leads us to William Gilbert, who published his *De magnete* in 1600. In fact, Gilbert's direct intellectual ancestor was Peter Peregrine who wrote, in 1269, a book of the same title, *De magnete*, which shows a remarkably modern attitude towards observation and experimentation. Some near-contemporaries, especially Cardan and della Porta also wrote about magnetism. Gilbert argued against Cardan who had asserted that magnets feed on iron;[89] and he debated della Porta who had stated that lodestones have hairs and that iron is 'intoxicated'.[90] Empirically, Gilbert benefited a great deal from two important Renaissance discoveries, the declination and the inclination. To the latter, Gilbert's idea that the earth is a huge lodestone and his *terrella* (small earth) model provided a beautiful explanation — by pivoting a needle across a *terrella*, Gilbert discovered that at the equator the needle had no dip, whereas at the poles it had a dip of 90 degrees.

More importantly, Gilbert changed scientific philosophy, as we have mentioned before, by denouncing the concept of 'sympathy' vs. 'antipathy', and by advocating the concept of physical forces. His pioneering work had significant influence among his scientific followers. One of his passages in this regard stated:[91] 'Wrongly therefore do those who discourse on these matters infer an antipathy (that is, the force of repelling by contrary passions); for there is not repelling force in them; and repulsion comes from action, not from passion.'

Galileo and Kepler became fervent admirers of Gilbert. Galileo was very aware of the iconoclastic character of *De magnete*. He used his spokesman Salviati to say: 'The book of Gilbert might never have come into my hands, if a Peripatetic philosopher of great fame, to free his library from contagion, as I believe, had not given it to me.' Kepler in fact turned Gilbert's 'magnetic philosophy' into his 'new astronomy'. Unlike his contemporary scholars, Gilbert was neither an encyclopaedist nor a practitioner. He experimented, systematised, and theorised in a specified field of terrestrial phenomena, and successfully. In this way he signified the renovation of physics. With his concept of physical force, speculation on cosmological attraction, and championship of the Copernican system, Gilbert broke the ground for theorising mechanics. On the other hand, magnetism could make no further progress unless mechanics could in turn aid it.

This brief presentation of the development of Western magnetism further strengthens our arguments about China's scientific stagnation. Magnetism, as Needham tells us, was traditional China's stongest branch of physics, and it had developed very early. Yet it did not surpass a qualitative description of magnetic declination. The richness and trailblazing role of *De magnete* provide a foil that accentuates contemporaneous Chinese negligence of, or insensitivity to, physics.

When Song Ying-xing printed his *Tian Gong Kai Wu* in 1637 he was distressingly aware of the unmarketability of his brainchild, a commonsensical and pictorial description of a series of homely technologies: agricultural, manufacturing, and military. If *De magnete* had been translated into Chinese early in the seventeenth century, what would Chinese literati have thought about it? It would have fared no better than Euclid's *Elements* did when the latter was first introduced into China. One implausible situation is that in thirteenth-century China some literati of a mathematical proclivity would have learned Euclidean geometry as mathematical games, and after being familiar with it, would have tried to apply it one way or another. It is even less likely that *De magnete* could have attracted a group of readers in seventeenth-century China, and produced any scientific influence which was in any sense comparable to what it did in Western Europe. Western Europe produced *De magnete* because it was the culture that produced a wide circle of potential readers, followers, and innovators.

The degree and character of intellectual activation and creativity on the two ends of Eurasia were very different. D. Price describes England at the time when Gilbert was working on his book in the following manner:[92] 'During the last two decades of the 16th century, professional surveyors, navigators, makers of mathematical instruments, and teachers of their uses became very numerous and enjoyed considerable status, especially in the metropolitan seaport of London.' At the same time, after 1582, Matteo Ricci began his energetic activities as 'missionary through astronomy' in China. The contrasting results were that Ricci, with his Jesuit comrades, Chinese followers, and successors, after 150 years of effort, could not make European science take root in China — because they changed nothing in China's general attitude towards academic activities. During the same period, Western Europe developed modern science and was prepared to reap abundantly the first intellectual, political, and technological fruits of the Scientific Revolution.

15 A Summary

This outline of the main features of the physical sciences in traditional China has, I believe, convincingly presented my theme — Chinese stagnation. Mechanics remained rudimentary. Astronomy had a long and continuous tradition of taking notes of 'yi-xiang (abnormality)', but remained empiricist for 3,000 years. Geometrical optics was a topic to be picked up and put off time and again. Mathematics had early glories in the recognition of the 'Chinese Pythagorean theorem', the evaluation of pi, and the analysis of indeterminate equations. But the notion of rigorous proof never appeared, and the necessary notational innovations never occurred. In general, mathematics remained in an incomplete, pre-calculus stage. Magnetic study also enjoyed some early glories, but long stagnated. The mathematical study of the musical scale made steady progress and reached a marvellous climax by the end of the sixteenth century. But even in this field the alienation of theory from practice was prominent — a situation that barred further Chinese musical progress and its positive influence on other fields.

For us the idea of stagnation emerged in the context of the growth of modern science. Hence the importance of the criterion of scientific 'modernity', which recognises mathematical mechanics as the leading field and the accomplishment of the 'Newtonian' synthesis, that is, the so-called 'classical' mechanics, as an objective yardstick. Sivin has criticised Sinologists who listed 'cultural factors' that limited the development of Chinese scientific thought, yet who never read 'the literature of science in the Chinese language'.[93] By 'the literature of science', as can be seen clearly from his paper 'Copernicus in China', he meant mostly the scientific and astronomical literature made by Jesuit missionaries and their students since the turn of the sixteenth and the seventeenth centuries. What he has elaborated in this and other papers is that Chinese astronomers, after studying the Jesuits' books, 'began using geometrical models and methods and relating them to questions of cosmology', that the most brilliant among them were critical and innovative, and that they could develop their cosmological models without running into personal risks — all of which amounts to saying that some Chinese astronomers were good students while the Jesuits generally could not act as good teachers and could not introduce the most up-to-date Western astronomy (Copernicanism). 'Would Copernicanism Have Been Acceptable to Chinese Astronomy?' There would not be much controversy about this question. Yet that does not mean seventeenth-century China was prepared for a scientific revolution, as Sivin's 1982 paper suggests.[94] By my criterion of scientific modernity, the actual condition of seventeenth-century China was a far cry from Sivin's optimism. By the macro-historical outcome, as shown in the condition of China in the following centuries, my conclusion is retrospectively correct — as I have mentioned in Section 10.

Nevertheless, the seventeenth-century astronomical achievements, or the so-called 'Chinese scientific revolution', illuminate some interesting questions. Wang, Mei, and Xue's mastery of Western astronomy and calendrical know-how, though brilliant, fell in a specialised realm. In a sense, their works embodied and anticipated the late nineteenth-century policy, or topic of debate: 'Chinese learning is our substance; take Western learning as applications.'[95] Traditional China had always demanded a sophisticated calendrical science, and since the thirteenth century, employing foreign experts (Indians and Arabs) had probably been an uninterrupted tradition. Jesuit astronomical introduction and the inheritance of it by Chinese astronomers continued that tradition, which had never posed a threat to the 'Chinese substance'. A few generations before Wang, Mei, and Xue, Matteo Ricci's close collaborator Xu Guang-qi said: 'Let us melt their [Western] material, and cast them into the mould of the traditional calendar.'[96] Mei Wen-ding's claim that the Westerners had learned Chinese mathematics sometime in high antiquity also reflects, in a way, the zeal to protect Chinese centralism, and ideologically, the 'Chinese substance'.

By logical inference the idea of a so-called 'Chinese scientific revolution' supports our theorem that 'software decides'. It is ironic, because nothing can be more remote from reality than the supposed occurrence of a 'Chinese scientific revolution' in the seventeenth century. This so-called 'revolution' made nothing truly different, except perhaps in a very small professional circle and for only a short period of time (since about the 1920s the leading astronomers of Chinese observatories were all returned students from the West). In the wording of the nineteenth-century policy that we have just referred to, the 'substance' is our 'software', and 'applications' are our 'hardware'. Calendrical and planetary astronomy constitute but one item in the wide range of 'hardwares'. If some items of 'hardwares' were modernised, or revolutionised, or more frankly, Westernised, so long as the 'substance' remained undisturbed, the country was still its old self: its general scientific condition could hardly be described as modernised. This was especially true in the past. Nowadays the basic domestic strategy of totalitarian nations is to keep their 'substance' — their political institutions, which is tightly combined with an ideological control — in a predetermined fashion, while trying hard to modernise their whole range of 'hardwares'. Science does not grow wholesomely under totalitarian rule.

16 Whence My 'Inertia' Topic?

The inertia topic must have been in my mind for a long time, if intuitively. All my life I have been taught that China stands among the longest and most civilised cultures; yet all my life I have been aware (and ashamed) of China's inferiority in terms of economy, popular literacy, scientific performance, military power, and political enlightenment. All my life I have resented wasting hour after hour to master a decent calligraphy, because I was aware that such formal exercises do not produce the substantive knowledge that is so desperately needed to rescue China. For a few decades I was enthusiastic about the Romanisation — 'pin-yin' — of the old Han characters, because I once believed that that reform might serve as a short cut to raise the cultural level of the people. When I was a young man, I put *Tang-shi San-bai Shou (Three Hundred Tang Poems)* and *Gu-wen Guan Zhi (The Pinnacle of Classical Prose)* alongside my music albums, because I thought such Chinese cultural heritages as these collections, though divinely beautiful in their own terms, had little to do with the useful insight I have sought. All my life I have been distressed by the knowledge that my ancestors were so preposterous as to insist on a set of cruel standards for females (Neo-Confucian 'chastity' since the thirteenth century, binding feet for over a millennium), and that in the eighteenth and nineteenth centuries my ancestors were so hypocritical and weakminded that they daily repeated such Confucian slogans as 'be loyal to the emperor and love China' while being lured to smoke opium!

However, it was not until I started reading Needham's works about three years ago that the scientific inertia topic occurred to me. (When in China, Needham's original books were not available to me. Before I left China the first half of Needham's first volume was published in a Chinese translation, from which I could not form any special ideas.) Owing to the nature of my early readings (*The Grand Titration* etc,), I first perceived my difference with Needham as terminological, and later as logical. That perception, though very incomplete, obliged me to go into a comparative study, which in turn formed the basis of a theoretical exposition (Essay Two, Chapter 3). Finally, after I had studied Needham's volumes I, II, III, and Part 1 of Volume IV, I knew that my deepest difference with the Needham school was philosophical. In Essay Three of this collection I discuss in detail the philosophical difference which centres on the pivotal 'law of physical causation'.

Certainly I dismiss neither the great significance of studying China's scientific past, nor Needham's voluminous contribution to this enterprise. Yet I must present my thinking on the inertia topic as clearly as possible. Needham has been anxious to change the misconception that China had nothing worthwhile to offer, and that China had been stagnant. (He was half right: China had plenty to offer, but none the less became stagnant.) I am more worried about China's general and persistent reluctance to respond positively to a rapidly changing world.

Finally I will supply another piece of 'phenomenological evidence' that 'demonstrates' the Chinese inertia. When discussing my paper, Bill (Professor Monter) suggested that I use Lord Acton's adage: 'We study the past in order to overcome it.' Since I urged him to contribute a foreword for me, I wanted him to use it. During the 'haggling', I said: 'I will try hard to find a Chinese equivalent. If I fail to do so, then the failure testifies to the Chinese inertia.' Bill agreed. Now several months have passed, whereas any Chinese saying of the kind that I could recall purports: learn from the past, and imitate it; or somewhat better, derive lessons from it — 'quian-shi bu wang, lhou-shi zhi shi (the memory of what had happened guides what follows)'; the monumental multi-volume history compiled in the eleventh century by Sima Guang was entitled *A Comprehensive Mirror That Assists Government (Zi Zhi Tong Jian)*. The ancients represent the good old model; saying *overcoming* the past is very impious and anti-gerontocratic in China.

Notes

1. I console myself with two pairs of propositions:

'Full-fledgedness' does not necessarily imply persuasiveness; conversely, persuasiveness does not necessarily require 'full-fledgedness'. Original historical research does not necessarily mean originality; conversely, originality in historical studies does not necessarily demand original research.

I am not sure how much originality I can claim, but I know that I owe many original researchers a great deal.

2. The debate around this hypothesis comprises part of the difference between Needham's faith in the organicist philosophy and my faith in the traditional 'law of physical causality', or mechanistic, or reductionist view. Therefore the debate is by no means trivial, as it is made clear elsewhere and in Essay Three (Chapter 4).

3. H. Butterfield, *The Origins of Modern Science* (The Macmillian Company, New York, 1962), p. 116: 'In the synthesis of Descartes . . . there is the economy and austerity of a highly concentrated deductive system. By its mechanisation it anticipated the structure that physical science was to assume in the future.'

4. J. Needham, *Science and Civilization in China* (Cambridge University Press, 1962), vol. IV, p. 1.

In the following, to avoid repetition, Needham and his collaborators' volumes in this series will be simply referred to by the number of the volume.

5. See, for example, *Encyclopaedia Britannica* (1975), vol. 15, p. 120a: Maxwell's 'great object, as it was also the great object of Faraday, was to overturn the idea of action at a distance'; and 'Maxwell showed how to reduce all electric and magnetic phenomena to stresses and motions of a material medium.'

6. A parallel question exists for the priority problem between Euclidean and non-Euclidean geometries. On the ground of pure mathematical logic, there is no necessity to exclude the precedence of a non-Euclidean geometry. Here, however, mathematical logic may differ greatly from historical logic. This interesting feature reflects the difference between physics and mathematics: the former deals directly with the real world (hence its development conforms with historical logic — simple things and things of immediacy first); the latter are essentially — or can be — logical games built on artificial rules; once it reached maturity, its logical developments might appear to be historically erratic.

7. As a Chinese I am determined to contribute my every effort to recover the glory of China's past civilisation. Criticism logically precedes a correct evaluation; my self-critical stand and 'the emperor is naked' shouting attitude are rooted in this determination.

8. Vol. II, p. 161.

9. Ibid., p. 163.

10. Graham and Sivin, 'A Systematic Approach to the Mohist Optics' in Nakayama and Sivin (eds), *Chinese Science* (The MIT Press, Cambridge, Mass., 1973), p. 115.

11. Vol. II, p. 281.

12. Graham and Sivin, note 10 above. See the sections on 'The Optical Propositions', 'Translation', and 'Problems of Comprehension and Translation'.

13. Vol. II, p. 304.

14. Ibid., p. 303.

15. Sivin, book review in *Scientific American*, vol. 226, no. 1 (Jan. 1972), p. 113.

16. A history of the introduction of Euclid's *Elements* into China — in the thirteenth century (probable), in 1607 (the first six books), in 1857, and in 1865 (interestingly enough, by the most prominent Confucian official of the time, Zen Guo-fan) — is outlined in Needham's vol. III, pp. 105–6:

> The evidence does not absolutely indicate that the books in question were a translation of Euclid into Chinese, though it seems probable that they were. In any case they had no perceptible effect, and during the following centuries the survey geometers, such as Tang Shun-zhi and Zhou Shu-xue in the +16th century, continued quite naturally to follow the ways of their predecessors in the Tang and before.

As late as in the 1860s Zen Guo-fan might recommend Euclid, as he had done for the establishment of shipyards and arsenals. But he would have never suggested the abolition of the examination system, which occurred 40 years later in 1905, after the humiliating Sino-Japanese War in 1894, the tragic 1898 Reform, and the disastrous Boxers' Rebellion in 1900. Zen Guo-fan's politico-ideological stand could have never surpassed 'Zhong-xue wei ti, Xi-xue wei yong — Chinese learning is our substance; take Western learning as applications.'

17. N. Sivin, 'Copernicus in China', *Studia Copernica, 6*, 1973, p. 64.

18. C.C. Gillispie (ed.), *Dictionary of Scientific Biography* (Charles Scribner's Sons, 1970), vol. II, p. 606.

19. Vol. IV, pp. 39–40.

20. Ibid., p. 39.

21. Ibid., p. 40.

22. Ibid., p. 39.

23. Ibid., p. 42.

24. Ibid.

25. The Institute of the History of Natural Science, Academia Sinica (eds), *Zhong-guo Gu-dai Ke-xue Ji-shu Cheng-jiu (Scientific and Technological Achievements in China's Past)* (Beijing 1978, 1980 reprint), p. 147.

To avoid repetition, the book will be referred to by its English acronym, STACP, in the following.

26. The English translation, which I rendered above, is based on the Chinese researcher Dai Nian-zu's explanation. Nowadays the majority of learned non-experts would find the archaic language too sparing and too obscure to follow. I suspect that learned people in the past also had to face up to similar difficulties — perhaps even greater difficulties in terms of a scientific understanding.

27. D. Joravsky, 'Scientists as Servants', *The New York Review of Books* (28 June 1979), pp. 37 & 38.

28. Vol. IV, p. 22.

29. Ibid., p. 23. Preceding this *Huan Nan Zi* passage Needham's argument is '. . . it shows that the Mohists must have been essentially in possession of the whole theory of equilibria as stated by Archimedes. That it was widely understood in the Han emerges from a passage in the *Huan Nan Zi*.'

30. Vol. II, p. 582.

31. STACP, p. 184.

32. S. Nakayama, 'Science and Technology in China' in A. Toynbee (ed.), *Half the World* (Thames and Hudson, London, 1973), p. 144.

33. Vol. III, p. 152.

34. Ibid., p. 103.

35. STACP, p. 104.

36. Vol. III, p. 101.

37. Fan Wen-lan, *Zhong-guo Tong-shi Jian-bian (An Outlined Chinese History)*, Vol. II (Ren-min Chu-ban-she, Beijing, 1957), pp. 444–5.

38. Vol. III: pp. 75–7. STACP, p. 79.

39. STACP; pp. 78–9 and p. 75; vol. III, p. 76.

40. STACP, p. 79.

41. Vol. III, p. 119.

42. STACP, p. 114.

43. Ibid., p. 132.

44. Vol. III, p. 154.

45. Ibid.

46. Ibid., p. 158.

47. Ibid., p. 154.

48. Ibid., p. 155.

49. When reviewing my manuscript Dr Robert Michaelson added a quotation to this 'not so respectable' observation of mine: 'While indeed Newton did not use the

notation of fluxions in the *Principia*, a modern mathematician, according little respect to those who confuse notations with notions, finds the *Principia* a book with the theory and application of the infinitesimal calculus.' (C. Truesdell, *Essays in the History of Mechanics*, p. 99.)

Our issue is: mathematisation or axiomatisation? And my conclusions, which I hope are more respectable, on this issue are:

Axiomatisation serves any field of explanatory sciences as a guiding principle, whereas mathematisation is less general;

While mathematisation may apply to phenomena of diverse nature, axiomatisation aims at establishing a unifying and explanatory science.

50. Vol. II, p. 151.

51. Vol. II, p. 136.

52. Vol. III, p. 167.

53. Ibid., p. 171.

54. Ibid., p. 460.

55. Ibid., p. 399.

56. Sivin, 'Shen Kua' in *Dictionary of Scientific Biography*, note 18 above, Vol. XII, p. 375a.

57. Ibid., p. 378a.

58. Sivin, 'Wang Hsi-shan', note 18 above, vol. XIV, p. 161b.

59. Sivin, 'Why the Scientific Revolution Did Not Take Place in China — Or Didn't It?' in *Chinese Science, 5* (June 1982), p. 62. See also 'Wang Hsi-shan', note 58 above, p. 160b.

60. Sivin, Wang Hsi-shan', note 58 above, p. 160b. The bracket was presumably inserted by Sivin.

61. Scientific — generally, historic-social — inertia was not unique to the Chinese character. But traditional China seemed particularly incapable of overcoming it, and that comprises our major comparative concern: How could Western Europe overcome it? Why could not traditional China overcome it?

62. Vol. III, p. 468.

63. Ibid., p. 480.

64. Ibid., pp. 485–7.

65. Ibid., p. 490.

66. Ibid., p. 492.

67. Ibid., pp. 470–1.

68. Ibid., p. 478.

69. The irrelevance of a weather vane to a meter pointer reminds us of another irrelevant connection, that is between the abacus and the coordinate system. (See vol. III, p. 107.) A coordinate system relates a point on a plane with an ordered pair of numbers. Even limiting to the one-dimensional case, 'a point' in an abacus is just one digit in a number. Sometimes superficial signs of similarity lead us to the misconception of having recognised true parallelism. Besides, this situation is not limited to the analogy between specific objects. In the case of the *Huai Nan Zi* 'theory' of the lever, or in Han technicians' mastery of 'Archimedes' principle', we were likewise confused. Chinese toiling masses, leading a subsistence life, had a saying: 'Zhua jin lan li bian-shi cai — Whatever is dumped into my basket will make me a vegetable dish.' Do we have to be so indiscriminating when dealing with Chinese scientific history?

70. Vol. III, p. 466.

71. STACP, pp. 293, 294, & 296.

72. Vol. III, p. 543.

73. Ibid., p. 570.

74. Ibid., p. 548 & Plate LXXXI.

75. Ibid., pp. 551–2.

76. Ibid., Plate XCI, facing p. 583.
77. Ibid., p. 634.
78. Ibid., pp. 624–5.
79. Ibid., p. 625.
80. Vol. IV, p. 212.
81. STACP, p. 164.
82. In C. A. Ronan, *The Shorter Science and Civilization in China* (Cambridge University Press, 1981), vol. 2, p. 385, Zhu is mistakenly identified as a 'Manchu prince'.
83. Vol. IV, p. 216.
84. Ibid., p. 224 and setion iv.
85. Vol. IV, pp. 249–50.
86. Ibid., pp. 302–3.
87. Ibid., p. 313.
88. Ibid., p. 314.
89. William Gilbert (ed. D. J. Price), *On the Magnet* (Basic Books Inc., New York, 1958), pp. 37 & 92.
90. Ibid., pp. 66 & 138.
91. Ibid., p. 113.
92. Ibid., p. vii.
93. Sivin, 'Copernicus in China', note 17 above, p. 75.
94. Note 58 above.
95. See note 16.
96. See for example, A. Toynbee (ed.), *Half the World* (Thames & Hudson, London, 1973), p. 132.

3 SCIENCE AND TECHNOLOGY IN TRADITIONAL CHINA: A SOURCE ITSELF INERT

1 Basic Chinese Dichotomy — Instability Within Continuity

Hegel's image of 'eternal China' connotes quiescence. But China could only appear quiescent to Europe at a time when communication was far more primitive than it is today. China's significant impact on world civilisation has never been flamboyant. It is true that the Han people gradually expanded their territory. Yet, more often than not Chinese expansion had been realised in ironic ways: the Han people themselves were first conquered, but then acquired new land in the course of driving their enemies away, or retaking their lost territory, or 'Hanised' their conquerors. Viewed from afar, China seemed to lead a stable and ponderous existence. Yet internally it was never quiet and stable. No other nation had so many peasant rebellions, was plagued by so many civil wars, and was invaded so often; yet no nation preserved its own characteristic culture so well. A paradoxical situation existed up to 1839: China was often too weak militarily, or politically, to resist invasion, but survived because of[1] 'the extraordinary integrative and absorptive power of Chinese civilization, a power which no invader before modern times was able to withstand.'

The existence of the unique Chinese 'dynastic cycles' indicates that there have been some dominant and persistent factors that sustain China's continuity. I think the major factors are:

(1) the early development of a high civilisation by the Han people, which was more advanced than that of the surrounding peoples;
(2) the strong ethical appeal (filial piety, extended family, equality among brothers, and inequality between males and females, etc.) and political persuasiveness (autocracy and hierarchism) of Confucianism in a millet- or rice-planting agrarian society;
(3) the vastness of Chinese population and territory.

With the aid of its celebrated (now notorious) system of civil service examinations, the dominant ideology of traditional China effectively supported and co-operated with a unified imperial rule. The resultant politico-ideological ruling body, dominating so large and relatively closed a subcontinent and during times of underdeveloped technologies, prevented China from breaking away from a stagnant status. The cardinal principles of the Confucian ideology maintain a huge population and its stubborn macrohistorical growth. Consequently, the Chinese people were caught between a two-sided Confucian pressure: overpopulation and dynastic autocracy. Except at the heights of the Qin-Han era, the Sui-Tang era, and the Ming Dynasty, the military power of the Han people has never been strong enough to spare them the fear and anguish of suffering from foreign invasions (even during the Han) or piratical harassments (even during the Ming). The narrow vista and inactivism of all leading Chinese philosophies smothered the Chinese people's initiatives and innovativeness for too long. The overwhelming heritage of a system that combined politico-military power and ethic-ideological authority makes the Chinese people vulnerable to totalitarian rule, which in a sense is prolonging the Chinese inertia. Although the Chinese people have been innovative in mechanical skills and technologies, traditional China's politico-ideological inhibitions kept the Chinese people from making direct contributions to the theoretical infrastructure and methodological foundations of modern science. The generalised dichotomy — instability within continuity — was the overall consequence of a number of phenomenological dichotomies: ideological strength (or rather, effective ideological control) vs. political (and economic) weakness; cultural integrativity vs. military vulnerability; capability of self-preservation vs. incapability of adequate self-defence; subsistence vs. adequate self-sufficiency; intellectual continuity vs. the lack of creativity; high cultural achievements in certain areas vs. the lack of comprehensive cultural progress.

2 Looking at Traditional China from England

Not only is it interesting, but increasingly a necessity to view a country from a foreign, or world-wide, or comparative perspective.

About 350 years ago, when Francis Bacon explained the historic impact that three great inventions (printing, gunpowder and the magnetic compass) had on the world, the Chinese people themselves would not have dreamed that they or, to be exact, their ancestors, merited such credit. On the one hand the Chinese had the self-complacent idea that everything civilised came from the Middle Kingdom; on the other hand, lacking a broad perspective, they could not evaluate the significance of their own creations. Since the 1930s and 1940s, another Englishman, Joseph Needham, started his lifelong study of the inventiveness of the medieval Chinese people, and he arrived at conclusions such as this:[2] 'Not only the three which Lord Bacon listed . . . but a hundred others — mechanical clockwork, the casting of iron, stirrups and efficient horse-harness . . . — all had their effects, sometimes earth-shaking effects, upon a Europe more socially unstable.' Since then Needham has repeatedly raised the question: With so many technological priorities, why did China not develop modern science and technology independently? Using his own words,[3] 'the more you know about Chinese civilization, the more odd it seems that modern science and technology did not develop there.' Needham identifies this question as[4] 'one of the greatest problems in the history of culture and civilization — namely, the great problem of why modern science and technology developed in Europe and not in Asia.' Sometimes he put this interesting macro-historical topic of comparative studies in more specific forms: Why did Chinese science and technology always remain primarily empirical? Why was there no indigenous industrial revolution in China?

3 The Needham Puzzle: Its Legitimacy

In 1973, after this question, which I have called the 'Needham Puzzle',[5] had been raised and explored for more than 30 years, A. C. Graham argued eloquently that it is in fact pointless to ask this 'negative question'. His paper, entitled 'China, Europe, and the Origin of Modern Science: Needham's *The Grand Titration*', was published in the collection dedicated to Needham's 70th birthday. Graham wrote,[6] ' "Why was there a Scientific Revolution in Europe about 1600?" and "Why was there no Scientific Revolution in China or India?" — questions that look as though they are two sides of one coin'; and[7] 'although the positive question is real and important there is something wrong with the negative question . . .'

When a pair of things are antithetical to each other as are the two sides of a coin, why is it wrong to raise questions about either of them? As for the present question concerning the emergence of modern science, why is it wrong to inquire 'why did not China succeed?' when the answer could very well be, 'because China then or today lacked such and such necessary conditions'. Is it not every bit as justifiable and meaningful as to ask 'why did the West succeed?' and arrive at a conclusion something like 'because it possessed such and such sufficient conditions'.

As a matter of fact, a few pages later, Graham told his readers, in effect, that his two questions were not the two sides of a coin. This time he wanted to emphasise that they simply did not form a coin: while there is one positive side, there are numerous negative sides. If you ask the negative question, he says, you will have to run through a list of numerous nations. That is what Graham meant by 'there is something wrong with the negative question'. He used analogies to illustrate his position: it is pointless to run over the list of maritime nations asking of each why its swimmers did not discover the crawl before its dissemination from the Polynesians.[8] (Another analogy of his is the invention of skis.)

Does asking a negative question necessarily mean running through a whole list? By no means. Was the invention of the crawl, which could very well be the achievement of a single swimmer, a good analogy for the emergence of modern science? By no means. The crux of the problem is that the works of Needham and others revealed more and more evidence of medieval China's leading position on many technological matters. Then, very naturally, indeed unavoidably, the question appeared. There is no necessity of running through any list; or at least the list is happily very short. China enjoyed the honour of being selected because she seemed such a logical candidate for the 'inventor' of modern science. Needham justified his questioning by his voluminous research. India was somtimes mentioned, but no other nations were picked at random. It was Francis Bacon who first pointed out the historic significance of the three great inventions which were later identified as Chinese contributions; Needham pushed one step further to ask: Why was China herself immune to her Europe-shaking inventions?

Now may we not carry forward this marvellous tradition in comparative study and inquire: What road-blocks impeded traditional China? Is it not meaningful to seek out these factors? Possibly they are still functioning today as inhibiting agents. Is it not of consequence to pinpoint them, particularly at a time when the Chinese people still have so much difficulty in transforming their motherland: when a developing nation imports and then implants modern science into her land, her people are 'answering' the negative question with their social and intellectual activities. In redressing the lag, people 'answer' the question of lag in practice. So, it is not pointless to try to answer it theoretically.[9] A theoretical answer can sometimes contribute substantially to a practical 'answer' — although, to be honest, this contribution might be either positive or negative.

4 The Undeniable Great Inertia

While raising his 'negative question', that is the Needham Puzzle, Needham criticises the use of words like inert, stagnant, or changeless to describe China. More than once he emphatically states that 'the word "stagnation" was never applicable to China at all; it was purely a Western misconception'.[10] To him, medieval China was almost at the threshold of the spontaneous development of modern science; certainly it was not inert. Instead, he likes another set of adjectives: homoeostatic, cybernetic,[11] that is, self-regulative. He wrote:[12]

> For there was something in Chinese society which continually tended to restore it to its original character (that of a bureaucratic feudalism) after all disturbances, whether these were caused by civil wars, foreign invasions, or inventions and discoveries.

In economics, such a state would be called a steady state. In classical physics, it is an equilibrium state, and in quantum physics, a 'ground state'. A marble in a bowl is the simplest case. If you look at the marble from inside the bowl, you find it tends to return to its original state or position after each disturbance. From the view outside the bowl, the marble-in-a-bowl system as a whole is a stationary, or 'inert', or 'stagnant' one, so long as the marble is confined by the bowl. Needham and I both look at the marble-in-a-bowl system. While I am waiting in vain for the marble to 'jump' out of the bowl, I say: 'What a stagnant system!' Needham, on the other hand, concentrates on the movement of the marble — with the oscillatory nature of the movement he is familiar — and says:

'How can you say that? Don't you see the marble moves unceasingly?'

The dichotomous characterisation presented at the beginning of this paper shows the general inertness of Chinese society: China had an early and continuous civilisation, but she also had a very difficult and much belated modernisation, including the non-emergence of modern science. According to this view, inertia in China is a fact, stark and undeniable, for either traditional China (*c.* 200 BC–AD 1839) or transitional China (1839–1945), or present-day (apparently again transitional, or neo-traditional) China. The question of whether China is inert or not should be decided by the evidence provided by the trends in *per capita* income, the degree of urbanisation, average literacy, and general and political enlightenment. With regard to a question like this, we have to rely on hard statistics, and on the net results of historical development. Ironically, the glorification of medieval China's inventiveness supports my argument. No matter how many early individual technological inventions there were, on the whole, they were not effective in pushing Chinese society ahead. Therefore, there must have been some very effective inhibitions that neutralised the positive influences of all these inventions.[13] Is it not plain logic to say that the more medieval China's inventiveness is documented, the more inertia we must attribute to her — in view of the undeniable final results?

In other words, when we contrast the Needham Puzzle, or the presupposition that backs it up, with the unsuccessful efforts at importing and implanting modern science into China[14] for the past four centuries (Matteo Ricci started working in China in 1582), one naturally asks: Had traditional China ever been a near-producer of modern science? If it had, then we present-day Chinese are confronted with a humiliating contrast between the alleged progressiveness of our ancestors and our own almost hopeless backwardness. Had there possibly been a process that transformed a forerunner into a straggler? Yes, quite possibly. According to Hartwell,[15] 'the establishment of a Neo-Confucian, classicist-moral-didactic orthodoxy might, in many ways, be viewed as marking the beginning of "Traditional China" '. This was the so-called post-Song decline. Historians generally noticed that the Yuan-Ming-Qing era saw an evident retrogression from the Medieval Flowering (the eighth to thirteenth centuries), or from 'modernistic' tendencies, such as political pragmatism and economic and intellectual pluralism. Although some evidence of economic growth of a capitalist nature in the Ming Dynasty was discovered,[16] the macro-historical fact is that during the three to four centuries between the Ming and the Opium War, traditional China remained characteristically pre-capitalist. Therefore we have come across a series of significant questions: Why could not the Medieval Flowering continue? Was Neo-Confucian classicist-moral-didactic orthodoxy an unavoidable reversal? What were the major factors that hindered Chinese capitalist development? I would group such questions under the category of the 'Chinese inertia mechanism'. To the establishment of Neo-Confucian orthodoxy I would answer the question affirmatively and explain it with the major and everlasting feature of traditional China: the symbiotic combination of political rule and ideological control.

5 Modern Science and Modernity

In the above, however, I have failed to distinguish between modern science and modernity in general. Historically, the emergence of modern science was an indispensable part of the growth of modernity in general. One essential ingredient of present-day modernity is the application of modern science to the realm of production and government activities. Traditional China was not a country of 'philosophers in rags'.[17] We certainly do not want to build a country of 'starving scientists and shivering engineers'. I mean to argue that poverty has been one of China's most enduring characteristics, and a backward economy can never support a truly advanced culture. Therefore, it seems justifiable to identify the growth of modernity in general as the necessary condition of the development of modern science and technology.

On second thought, however, one finds that it is not easy to tell whether the growth of modern science and technology necessitates general and comprehensive modernity. If we include democracy (or modern politics) as one of the 'legs' of the 'tripod of modernity' — industrialisation (modern economy) and science (modern culture) being the other two 'legs' — one may cite a number of totalitarian states, past and present, which have achieved modern science and technology or are working hard toward it, as obvious counter examples. In other words, a 'biped modernisation', that is, modernity without democracy, is possible. Moreover, as Professor

Joravsky often explained to me, historically speaking, to say that modern science requires democracy is to slip over the entire history of science before the nineteenth century.

I was so inclined to support a theory that a modern system of knowledge (science), a modern system of economy (capitalism), and a modern system of government (democracy) were causally correlated, and more or less concurrent, historic developments. But it seems that my intuitive conclusion which was based on my present-day political concern has to yield to historical facts. The 'idea of concurrence' is one of the few Needhamite hypotheses which I supported for two years, but now I have to abandon it with a better recognition of European history. We will return to this question in Section 13.

Granting that there were no necessary causal relations between modern democracy and the economic and scientific aspects of modernity, some of my observations of the modern world may still retain some worth. The case of 'developed' totalitarian states is a historical phenomenon that occurs at the mature stage of natural sciences and engineering technologies — a situation that enables careerists, backed by aggressive or vindictive political philosophies, to deveop 'biped modernisation', which is characterised by numerous inherent self-contradictions and everyday hypocritical assertions on the part of the totalitarian authorities. In other words, 'biped modernisation' is an unnatural modern development; 'tripod modernisation' is the result, or the process, of natural and earlier historical development.

I may have to answer the elusive question: what are 'natural' historical developments and what are 'unnatural' historical developments? A tentative definition is this: historical developments which are heavily influenced by the personal will of political dictators are unnatural — if it were not for the dictator's control, the course of events would evolve differently. (I believe any person who has lived in Maoist China or who is adequately knowledgeable about it would agree that this last statement is not at all elusive, but tangible.) Returning to the question of modernity, I have recently become inclined to include 'population pattern' — its growth, structure, and quality — as an independent 'leg' and thus replace the 'tripod' by a 'quadruped'. Modernity should be thought of as a living force; in today's world, however, modernity as a whole is far from being guided by rational, let alone scientific, deliberation — thus a 'quadruped' serves as a good metaphor.

6 Foundation of Science: Einstein's Axiomatisation

To support his argument that raising the negative question is pointless, Graham also quoted in his paper the full text of a short letter of Albert Einstein's, which says that the foundations of science are formal logic and experimentation. Einstein also suggested that we had better not ask why the Chinese sages did not arrive at them, because the achievement of these two great discoveries was almost accidental, or providential. 'In my opinion,' Einstein wrote, 'one has not to be astonished that the Chinese sages have not made these steps. The astonishing thing is that these discoveries were made at all.'[18]

It is obvious that what Einstein wanted to stress is the serendipity of these two great methodological achievements. China was mentioned by him in passing because Chinese civilisation was one of the four earliest and the one which had the least to do with later Western development. I think Einstein's letter is interesting because he reduced, or axiomatised, the historical basis of modern science. Others have proposed other parallel sets of axioms. However, I do not take Einstein's letter too seriously, since he knew little about China, and since his almost facetious statement of the near-impossibility of the two great discoveries forestalls any possible curiosity about instances of non-development of science.

The basis of modern science was not laid by certain individuals, not by a few accidental discoveries or providential revelations, and not in a short span of time, but by collective or interactive efforts of groups of people, generation after generation. When such efforts were carried out consciously on a societal scale, and were supported by persistent incentives, modern science, including all its 'astonishing things', became possible. From the Industrial Revolution the British people arrived at the adage: 'Necessity is the mother of invention.' Every individual's necessity does not guarantee his or her becoming an inventor; hence necessity has to be understood as social necessity, that is, as a persistent incentive operating on a societal scale. I remember that the dolphin stroke, called the butterfly today, was an innovative technique invented by a Hungarian swimmer in the 1950s. We, however, can never talk this way about the emergence of modern science, for which we must adopt a philosophy roughly parallel to the above British conclusion, namely that social necessity is the mother of technological progress.

7 A Pure Materialistic Explanation?

As we have indicated before, by and large China has been a country of poor peasants; her average cultural achievements have, in fact, been limited. It is widely acknowledged that general poverty, a very difficult writing system, and the all-pervasive politico-ideological controls share the responsibility in this regard. The Han characters, that is the 'square words', which serve as the major conveyance of Chinese civilisation, are evidently deficient: they are too cumbersome a tool.[18] Consequently, intellectual performance facilitated by this writing system, in terms of net output, has been substantially reduced. Needham put forward a hypothetical statement:[20]

> It may not be too much to say that had the environmental conditions been reversed as between Euro-America and China, all else would have been reversed too . . . Westerners would have to learn the ideographic script just as the Chinese now have to learn alphabetic languages because the bulk of modern scientific literature is written in them.

In view of the point I have just expounded — that the net output of a tool is inversely proportional to its clumsiness, and taking into account the essential difference in the two systems of writing, one is tempted to offer an almost reversed hypothetical statement: Had the writing systems been reversed between Europe and China, Westerners would have had to learn science and technology from China as the Chinese now have to learn science and technology from Westerners.

I do not mean, however, to neglect the other 'environmental conditions'. For instance, I do not propose a simplistic theorem that reads: the owner of a better writing system leads in general culture, hence in science and technology. On the contrary, looking for basic causes has become my primary interest. I think I have finally located the root evil that reduces all other Chinese inhibitions, including the written language, to the status of minor culprits. No doubt, my original profession as a theoretical physicist moves me towards reductionism: that I now blame all problems on politico-ideological repression has become an unexpected axiomatisation. On the other hand, I certainly do not claim any originality in doing so. As is well known, 'Down with Confucius and Company' was the major call to arms in the agitated China of 1919.

In Needham's paper just quoted above, by looking for 'environmental conditions' he meant to establish a materialistic explanation of history:[21]

> The key probably lies in the four factors: geographical, hydrological, social and economic. All explanations in terms of the dominance of Confucian philosophy, for instance, may be ruled out at the start, for they only invite the further question, why was Chinese civilization such that Confucian philosophy did dominate.

Is this a meaningful way to pose a workable question? I doubt it. In this modern era of rapid change, or rapid industrialisation throughout the 'Third World' in particular, the roles played by 'geographical, hydrological, social and economic factors' are increasingly subject to suspicion. Not only are natural conditions too constant a variable, hence an invariable that fails to explain rapid changes, but also many find that social and economic factors are mainly acting as 'functions', not as 'independent variables'. To narrow the subject somewhat, I should say that I have increasingly become disillusioned about the possibility of ever finding some 'materialistic' explanations for some particular political institutions, historical personages, or extraordinary events which have played great roles in shaping the historical course of a nation or even of that world. Was the system of civil service examinations in traditional China instituted because of China's geographical, hydrological, social and economic factors? Were Mao Ze-dong's thinking and behaviour completely determined by his being a Hunanese, his family background, and his education? During the founding stage of Tokugawa Japan, both the so-called sankin-kotai (alternate attendance) system and the compulsory separation of samurai (aristocracy) from land were instituted, which profoundly influenced Japan's later urbanisation, hence modernisation. Was this situation the result of Japan's geographical hydrological, social and economic factors? I believe C. Totman has summarised Japan's significant step toward urbanisation in a succinct way:[22]

> One could even argue that the urbanization of early modern Japan was a bizarre and unintended byproduct of the countrywide application of two simple political practices of late sixteenth-century samurai leaders. One was the practice of

> holding hostages to assure the good behavior of one's allies; the other, the practice of requiring one's vassals to reside at one's castle.

No doubt, for my present argument, the key words in this quotation are 'a bizarre and unintended byproduct'.

8 The Basis of the Growth of Modern Science

In the final analysis, the early growth of science — which, as Sivin clarified recently, was independent of the growth of technology until almost the mid-nineteenth century[23] — depends on the urge for explanations of phenomena, the seeking for causes behind effects, and on the reasoned organisation of explanations conceived according to one's belief and perception of rationality. Therefore, from contemplating our basic question we turn to the following series of questions:

> Were scholars interested in natural phenomena?
> Did they look for rational explanations?
> Did social values encourage them to make persistent efforts in this direction?
> What was the attitude of the ruling politico-ideological entity towards scholars' efforts in this direction?
> Or rather, in a given politico-ideological environment, were there possibilities for scholars to develop their intellectual interests in these directions?

Because of inhibitions on and discouragement of independent thinking, as well as the greatly limited scope of scholarship in traditional China, the answers to these questions for China were increasingly negative. On the other hand, the significance of these five questions is in the order of their appearance increasingly decisive for the emergence of modern science. No doubt, this understanding is consistent with the proposition that modern science had to emerge together with a certain degree of political tolerance and enlightenment.

After more than half a century, it seems as if the premise of an interesting and provocative paper by Feng You-lan[24] 'Why China Has No Science?'[25] has been refuted for good. As S. J. Bennet describes the situation:[26]

> The demise of the 'China-had-no-science' myth has been chronicled by the monumental research project of Joseph Needham, *Science and Civilization in China*, and the coffin nails have since been driven home by the sophisticated studies of Nathan Sivin and Manfred Porkert.

In retrospect, this is indeed a natural stride forward. Feng You-lan stated clearly in his paper, 'By science I mean the systematic knowledge of natural phenomena and of the relations between them.'[27] He thought — probably by the standard of *modern* science — this systematic knowledge was absent in China, but research since than has disclosed that in traditional China there was 'systematic knowledge of natural phenomena'. Thus we have witnessed a cycle of progress: on the basis of a comparative study of Chinese and Western philosophies, a hypothesis was raised; this hypothesis was refuted by research into Chinese scientific and technological practices. Now we are pioneering a new cycle of progress, the starting point of which may be stated this way: we recognise[28] that a systematic (such as taxonomic and calendrical) ordering of natural phenomena such as existed in traditional China was not adequate for the emergence of modern science, which demanded progressive levels of explanations that finally, in the form of Newtonian mechanics, led to an axiomatic and quantitative comprehension. Traditional China did have natural science on a phenomenological level; note Max Weber's insight, 'Chinese thought has remained rather stuck in the pictorial and descriptive'[29] and Needham's observation that Chinese science has 'always remained primarily empirical'. The founding of modern science in the West, however, was the result of pursuing mathematically logical explanations for observed phenomena.

Before the European Renaissance, the age-old difference between the dominant and sophisticated Ptolemaic geocentric system and the ancient and primary heliocentric system was like the childless marriage of an inharmonious and unequal couple. As soon as the heliocentric part asserted itself, the marriage was consummated. The process that spanned the era of Copernicus, Galileo, Kepler, and Newton represents the labour pains that gave birth to a vigorous child. It is understandable *a posteriori* that the first branch of modern science, or the first grand unification, mechanics, was also the simplest case that could be put into a mathematical model.

9 What Keeps China Back — Two Chains of Logic to Answer

I believe millions of Chinese would today still find Feng's basic inquiry, 'What keeps China back?'[30] fascinating, and imperative. Sixty years ago, Feng's research led him to this chain of logic:

'What keeps China back is that she has no science.'[31]

Why no science?

'China has no science, because according to her own standard of value she does not need any.'[32]

Why did she not need any science?

'. . . China . . . devoted all her spiritual energy . . . to find good and happiness directly in the human mind . . . China tried to know what is within ourselves, and to find there perpetual peace.'[33]

To Feng, the answer to our question is purely philosophical: if Confucianism had not suppressed people's interest in the material world, if instead the people were encouraged to procure material from nature, the desire and necessity to control nature would have arisen, leading both to a Baconian quest for power and a Cartesian quest for certainty. Then, 'it is very likely that China would have produced science at a somewhat early time'.[34]

Now let us venture another chain of logic:

Why did modern science not appear in traditional China?

Because in traditional China intelligence and manpower were not activated and directed to rationalistic inquiries about nature.

Why not?

Because the politico-ideological entity in traditional China officially discouraged, hence socially despised, inquiries about nature and intellectual interest in technologies.

Again why? Or, why did this politico-ideological entity of traditional China function differently from the ruling bodies of Renaissance Europe?

In traditional China, a territorially unified autocratic rule was effectively aided by and symbiotically combined with an equally unified system of ideological control. Its philosophical spirit was introspective, its academic scope was officially limited and exclusively politico-ethical, and its basic attitude discouraged innovative practices and rationalistic inquiries. This politico-ideological situation contrasted sharply with European pluralism, which was fully embodied in its feudal separatism, national rivalries, and religious disputes. At the uppermost level stood the state-church bifurcation. European politico-economic dynamism and intellectual eclecticism had their root, or at least their potentiality, in this pluralism. In addition, other important facts must not be neglected.

First, in comparison with Confucianism, the Judaeo-Christian tradition is characterised by spiritual egalitarianism. As a result, individualism has been a much stronger spiritual heritage in Europe, and modern intellectuality demands independence. Second, Christianity views nature as the design of God. From the thirteenth century onwards, this notion was further supported by the idea of divine clockwork. Meanwhile, Aristotelianism was incorporated into theological argumentation and eventually extended to cosmological inquiries, a development regarded as pious fulfilment of religious obligation by many learned men down to the seventeenth century. Third, the great Greek heritage was extant, if dormant. But one may argue that in comparison with the necessary socio-historical conditions, the last factor was rather incidental (in conformity with my theory that 'software decides'): the invaluable intellectual legacy had lain untouched for centuries before the Western interest in it started to rise with the twelfth-century translation movement.

Therefore, the cause of intellectual hindrance in Feng's theory — the introspective Confucian values that delayed China's material and technological progress — is one of our corollaries. Feng proposed a philosophical axiom; we now offer a political, or politico-ideological, one.

10 An 'Independent Authority to Formulate the Laws of Nature?'

Before we taken an overall look at the dominant ideology in traditional China in terms of its historical inhibitions, we must confront a very interesting remark made by Sivin in 1979:[35]

> It is a simple fallacy to assume that the anti-scientific attitude of 'Confucian' humanists could have ruled out the passage to a mathematical physics of modern type. It is not at the center that revolution is generated, either in politics or in science. The

> fallacy becomes obvious when we note the enormous similarities between such negative attitudes of average intellectuals in China and those of the Schoolmen of Galileo's and Descartes' time. The Schoolmen could delay the dissemination of the new sciences, but could not block them or create alternatives as intellectually compelling. The Scientific Revolution held the initiative. It proceeded past them by constructing a new intellectual community (and careers for its members) outside the establishment of school and church, assuming independent authority to formulate the laws of nature, and displacing the prerogatives of ancient institutions in order gradually to form what has now become a technical establishment.

In spite of the 'enormous similarities', we have to note an important, indeed decisive, dissimilarity. In both China and Western Europe, emerging scientific attitudes were confronted with prescientific attitudes. In both subcontinents, antagonistic forces grappled in fierce struggles. But note well that they were on different topics, in fact in different fields, or at different levels. In China the study of nature, with few exceptions, was authoritatively denied and socially despised. In Europe, both the new scientists and the schoolmen were debating questions of cosmological structures and terrestrial mechanics, the interest in which goes back either to Greek achievements or to Christian tradition. In fact, Sivin knows well the contrasting Chinese situation: in one of his 1980 publications[36] he told us that Song Ying-xing, the author of *Tian Gong Kai Wu* (a book on Chinese technology written in the early seventeenth century) wrote angrily: 'Let the ambitious scholar toss this book onto his desk and give it no further thought; it is a work that is in no way concerned with the art of advancement in officialdom.' In the 2,000 years' history of traditional China, there were only limited geniuses like Song Ying-xing who shared a rebellious attitude; on the other hand, there were millions of 'ambitious scholars' whose major concern was 'the art of advancement in officialdom'. No socially influential 'initiative' on the part of people like Song, no 'new intellectual community and careers', no 'independent authority', and no new 'establishment' were possible in traditional China. The major cause was the official monopoly of a single system of ideology — a system which, to make the situation worse, was identified with official scholarship.

11 A List of Ideological Inhibitions

'Heaven changeth not, likewise the Tao changeth not.'[37] As a basic doctrine, this was summarised by Dong-Zhong-shu in the Former Han, that is, about 100 years before the Christian era. Dong, a well-known prime minister under the powerful emperor Han Wu-di, initiated the epoch-making slogan: 'Abolish all One Hundred Schools; advocate Confucianism only.' The content of Confucianism evolved in the course of history. Many elements of Buddhism and Taoism were absorbed by it to form an identifiable dominant ideology (and scholarship), which retained the name 'ru-xue' or 'ru-jia', Confucianism, later Cheng-Zhu (or Song-Ming) li-xue, the Chinese original of Neo-Confucianism. For 2,000 years China was shackled by this ideology of hierarchism, which explicitly endorses an obedient attitude, a self-contented outlook, and a pacifistic stand to deal with an allegedly changeless world.

J. Spence once described a tragic yet ironic episode which took place only a century ago during the declining years of the Qing Dynasty.[38] A Qing official, Wu Ko-tu, decided to protest against the policies of Dowager Ci Xi by taking his own life. In his last will he suggested that we Chinese should not accept Western things — from new gadgets to value concepts. Yet a few hours later, before he took his life, he wrote to his son asking to bring his photo — no doubt a very novel and Western artefact at that time — back to his home as a proxy of himself.

Both Needham and Fairbank described feudalism as the preparatory stage that led to capitalism and European modernity, and both authors think traditional China's lack of feudalism was a negative feature inhibiting her development. One surmises that the effective ideological aid of Confucianism prevented a return to feudalism after the Emperor of Qin's unification of China. Confucius' life-long call to arms was 'ke-ji; fu li — Restrain oneself; reverse to the (Zhou) ceremony.' He hated the chaotic (but not necessarily unproductive) separatism of the warring states. By this slogan he warned the state princes and advocated a return to the Zhou unification. Confucius was not a successful politician during his lifetime, but he was too successful as a political ideologist posthumously (much like Plato among Greeks).

For the greater part of his life, Confucius had difficulty in obtaining an official position in one or another court of the feudal states. His belief in the necessity of working under a superior ruler was so strong that he told his followers that he 'would be in great perturbation if for three months he did not live under the rule of a king'. He looked down upon efforts to earn one's material livelihood. In other words, he did not like to maintain his own welfare apart from a ruler's financial sponsorship. That intellectuals should choose to assert no independence at all appeared to be so natural in Chinese consciousness that as late as 1957, during the notorious 'anti-rightist campaign', people quickly 'accepted' the authorised 'theory' that 'if the skin does not exist, what can the hair adhere to?' Evidently, hair was a metaphor for the intelligentsia, and skin for the ruling class of society.

For centuries, to become literati-bureaucrats was the intellectuals' sole aspiration. Merchants were put at the bottom of the social scale ideologically: *shi, nong, gong, shang* (scholar-official, farmer-peasant, craftsman, and merchant). A merchant would try hard to bring up his sons along the dignified path of scholarship-officialdom, but never were offspring of bureaucrats expected to engage in trade as their first career choice. A folksaying went like this: 'Ten thousand occupations are low-graded, only scholarship ranks high.' Every year, in the long history of China, no matter which dynasty and which emperor ruled in the national capital, thousands of male candidates, young or no longer young, the great majority coming from families of landed gentry, worked hard in a narrow range of classical works, committed huge amounts of material to rote memory, struggled to master a cumbersome system of writing, engaged in fierce competition among their peers, all in order to obtain offices, whose credo was nothing but absolute loyalty to the imperial family. (The translation of the peculiarly Chinese term, *yu-zhong*, is not 'foolish loyalty', but 'as loyal as a fool' — a commendatory, not a derogatory phrase.)

In the history of Western Europe, we see that the emergence of modern science, that is the Scientific Revolution, was accompanied by the rapid assertion of the respectability of intellectuals (and merchants) and their unorthodox pursuits, and by mounting tensions between the State on the one hand, and the major source of ideological indoctrination — the Church — on the other. Compared with China, therefore, European society was not only ethnically, culturally, and religiously diverse, but also structurally bifurcated at the top level. These characteristics produced a very different politico-ideological system indeed, the evolution of which led finally to religious, intellectual, and political conflicts, which in turn forced upon an exhausted people the concepts of enlightened toleration and heterogeneity.

Fairbank had a remarkable perception of Confucian hypocrisy, though he did not characterise it as such. To use a Chinese phrase, it is an example of 'saying yes but meaning no'. After relating a 1936 episode and quoting a 1945 passage, Fairbank continued:[39]

> These two examples could be multiplied. They demonstrate the degree to which China down to recent decades remained a Confucian state. In the first case no one wanted Jiang to resign, nor did he intend to do so, and his resignation was elaborately declined. In the second case no one expected that China's national salvation in the midst of Japanese aggression, blockade, and inflation could be achieved through moral qualities alone, nor did Jiang think so. But in both cases his words delineated the traditional Confucian way.

How 'traditional' hypocrisy has been a 'Confucian' way I do not know. (Because Confucius himself included honesty in his eight moral principles, studying how hypocrisy finally became an acknowledged 'Confucian' trait in traditional Chinese society should make an interesting topic of research.) Since Mao and his party also display so much nauseating hypocrisy, I would guess it is more a Chinese than a bona fide Confucian trait. On the other hand, I am aware that 'Chinese' and 'Confucian' have been accepted by many people, and with plenty of reason, as interchangeable.

Confucian hypocrisy was especially manifest in its prohibition of communications between the opposite sexes and in its austere avoidance of nudity. The honest acknowledgement in Greek tradition of the beauty of healthy human bodies and the frank exhibition of it were absolutely inconceivable for Confucianists. (I do not know the ending of the 'mural polemic' over some innocent paintings in the Beijing International Airport which were finished in the so-called 'liberalisation movement' after Mao's death and the subsequent arrest of his wife in the fall of 1976.) In China's history we have never had a period that corresponds to the remarkably successful study of human bodies, from both medical and artistic standpoints, that occurred during the European Renaissance.

Hypocrisy as a deterrent to the growth of science is not limited to concrete cases. I propose using a 'Spencer gauge', after Herbert Spencer's idea of general evolution and his advocacy of the pre-eminence of science, to 'measure' the maturity of a system of thought, in particular, its degree of objectivity or subjectivity (superstition, hypocrisy, unverifiable and unverified assertions, *non sequitur* 'logic', etc.). By this standard, traditional Chinese medicine scores low. Lu Xun (1881–1936), the great Chinese writer, decided to study medicine when he was young because he had been so outraged by traditional Chinese medicine: among other things, a doctor had prescribed the hide of a broken drum to 'cure' his father's swollen stomach!

One can find evidence in present-day China to support the idea that the Confucian inordinate stress on filial piety and the 'logically' related male dominance formed the major cause of China's 'resilient' and macro-historically ever-growing population pattern. Less than a generation ago, Mao, when finally aware to some extent of the seriousness of the population problem, mentioned in a speech the case of a married couple who decided to add a tenth baby to their nine daughters.

One of the Confucian Cardinal Relationships is between brothers, one of respect and love, which upholds the eldest brother's position in case of the absence of their father. On the other hand, the Confucian standard of 'benevolence' obviously discourages the idea of primogeniture. As the population grew, a typical Chinese farm was divided into smaller and smaller plots. A Chinese folksaying partly reflects this fact: 'A family fortune does not last through three generations.' Meanwhile Chinese society could not derive benefits in terms of economic vigour and variety from primogeniture which, as evidenced in English and Japanese history, forced ambitious younger brothers to engage in different ventures.

China's arable land is very limited. Traditional ethics discourages any enterprise outside one's home town. In fact, an ideal Confucian society has only two castes: peasantry and bureaucrat-gentry. Although Chinese reality has never approached this idea, China's economy has been characterised by mediocrity and monotony, or to use R. H. Tawney's observation,[40] a maximum population competing for a minimum of resources.

Many have noted the Confucian contempt for manual labour. It would not be controversial to say that such contempt kept scholarship apart from practice. However, the major cause of this separation was the officially petrified scope and content of Confucian scholarship. We know how well the civil service examination system served dynastic China. It was the combination of this 'democratic' practice with an authoritarian ideology that produced remarkable effectiveness in dynastic rule. However, it was all too effective in limiting the vitality of academic pursuits. All other subjects were neglected or despised. Astronomy presents an interesting case, in fact an exception that proves the rule. Although every dynastic court definitely needed it, owing to the politico-ethical nature of Confucian scholarship, from the thirteenth century onwards traditional China could not produce enough competent astronomers for itself and dependence upon foreign experts became a tradition. Concerning the general character of Chinese astronomy, Kioshi Yabuuti wrote,[41] 'Even astronomy was not too abstract to escape being decisively formed by its political applications.' While Needham notes approvingly,[42] 'there were not people in the world who observed the changes in the heaven more persistently, and recorded them with greater amplitude and accuracy than the Chinese.' Yabuuti characterises the same practice, besides research in calendrical science, as 'unremitting observation of astrological omens'.[43] He also wrote:[44] 'Chinese astronomers were on the whole bureaucrats first, following established routines to discharge established responsibilities, and only secondarily researchers. It is natural enough that conservatism inhibited concern with new problems or new approaches to old ones.'

12 Some Illustrations of Sino-Western Contrasts

The 'outlook' of imperial-Confucian ideology was self-centred and self-complacent. Both curiosity (or ambition) in the exploration of things and places new, and acceptability (or tolerance) of change and variety, were limited by this outlook, or 'inlook'. Let us examine the Qing Dynasty (1644–1912), which should be the most interesting period for modern China. In the context of the study of traditional China and Chinese continuity, it was the last genuine part of Chinese traditionalism. The 'Hanisation' or Confucianisation of the Qing court was so thorough that after Kang XI (reigned 1662–1722) only a few things remained to remind people that it was a Manchu rule. What Emperor Qian Long (reigned 1736–96), 'yi-dai ying-zhu' (a wise lord), said and did were what could have been expected of a Confucian ruler. In 1793 Qian Long responded to George III of England with an edict that reads in part:[45]

> The Celestial Court has pacified and possessed the territory within the four seas. Its sole aim is to do its utmost to achieve good government and to manage political affairs . . . As a matter of fact, the virtue and prestige of the Celestial Dynasty having spread far and wide, the kings of myriad nations come by land and sea with all sorts of precious things . . . We have never set much store on strange or ingenious objects, nor do we need any more of your country's manufactures . . .

Pacification and possession of new territory, as a geographic continuation of China proper, was an alluring enterprise to which Qian Long himself contributed a great deal. The 'sole aim' was 'to achieve good government'. This is typically Confucian: the only concern of the imperial officialdom was the maintenance of standard ethic-political relationships, not the exploration of possible new relationships with nature, or with distant peoples — which is what Westerners had been busy doing for several centuries. 'Myriad kings come by land and sea' exemplifies the complacency of the 'Middle Kingdom'. Moreover, this condescending statement smacks of Confucian disdain of commercialism and contempt towards material acquisitiveness, an attitude which was traditional; although at this stage of Sino-Western contact, this 'leave-us-alone' and 'we-don't-need-sour-grapes' edict might already have been the result of a sense of inferiority, since China lacked such 'ingenious objects' and 'manufactures'.

In Europe three and a half centuries before, Henry the Navigator (1394–1460), the ruler of a small sea-faring nation, Portugal, had a very different frame of mind. There is no question that it was because of his vision that Portuguese mariners accomplished great deeds in oceanic explorations. Among other things they reached China around 1514, and landed in Japan in 1543.

Even earlier than Henry the Navigator, the Chinese eunuch Admiral Zheng He made seven voyages as far as eastern Africa between 1405 and 1430, with fleets composed of as many as 62 vessels and 37,000 soldiers. But right here we see some significant Sino-Western contrasts. First, the Chinese aim was not profit-seeking — or in any case, they were very incompetent profit-makers, and hence such ventures could not be self-sustaining. Rather to borrow Qian Long's words, it was to spread far and wide (Confucian) virtues and the prestige of the Celestial Dynasty. In the Ming Court the enterprise finally incurred severe criticism as a sheer waste and was stopped for good. Secondly, there was only one Chinese court that sponsored the enterprise, and there was only one Zheng He who repeatedly made arduous voyages. (In making these comparisons, we bear in mind that Ming China was about the same size as Western Europe, both in territory and in population.) Toward the end of the fifteenth century, when Western Europe became enthusiastic for oceanic exploration, it was not just one Queen Isabella who provided financial support and it was not just one Christopher Columbus who made great discoveries. When we look at the situation this way we perceive a big difference between imperially unified China and feudally, nationally, and religiously divided Europe. In order to survive, and to compete against its peers, every independent political unit had to mobilise its own manpower and intelligence to its full capacity.

The macro-historical paradox with Zheng He's pageantry was that it signalled the termination of Chinese naval power. The Chinese navy was rebuilt during the Southern (Later) Song Dynasty (1127–1279),[46] then it was used in the Yuan (the Mongolian) to try to conquer Japan (in 1274, Korean ships were used; in 1281, a combined fleet of Chinese and Korean ships were used). In the early decades of the Ming Dynastry (1368–1644), Chinese sea power continued to grow. But isn't it ironic that the apogee of the Chinese navy also marked the beginning of its end? Especially when the need for it grew, with attacks in the sixteenth century by pirates, Japanese raiders, and Portuguese navigators? Ray Huang thinks:[47] 'It was extremely paradoxical that in the middle of the 16th century the Japanese could violate China's security on the eastern seaboard. In many respects it seemed that the invasion should have been reversed.' We can now understand this paradox as the outcome of China's voluntary termination of its own naval power — owing to an 'increasing inward-looking defensive "fortress mentality" ',[48] which I identify with the traditional Chinese ideology.

Haven't we seen here another good illustration of the theory that 'software decides'? Zheng He had marvellous navigational 'hardwares', but the Ming court was the decision maker, whose way of thinking and political behaviour were in turn dictated by the dominant ideology.

Another meaningful Sino-Western contrast relates to the growth of universities. As early as the tenth and eleventh centuries, schools of medicine and law appeared in Italy (medicine in the south, law in the north). In the following few centuries, in almost all major cities of Western Europe universities prospered. Many of them, particularly the early universities in Italy — Reggio, Modena, Vicenza, Padua, and Vercelli — arose because they were independent of the civil and papal authorities.[49]

13 The 'Concurrence' of Three Great Historic Developments

As we have mentioned before, Needham hypothesised about the concurrence of three historic developments: capitalism, democracy, and modern science. To quote him:[50] 'There can be no doubt that capitalism, democracy, and modern science grew up together. The question arises whether the two latter are essentially dependent upon the former.' In Section 5 of this chapter, I confessed that I was much inclined to support this theory because I am a campaigner for Chinese democracy. Besides, I had an historical reason. That is, from our Chinese perspective, namely looking at the 'other side of the coin', Confucian inhibitions did not feature particularity, but generality. In other words, Confucian inhibitions on the development of democracy, capitalism, and modern sciences appear to be correlated. (Although admittedly, much analytic and synthetic arguments are needed to confirm, or to deny, such a conclusion.) But, when we switch to the European case, the crucial logic is that, even if this observation on Chinese history is correct, no one can derive from it the conclusion that in early modern Europe capitalism, democracy, and modern science must be causally linked.

I also mentioned before that I have been warned that we should avoid oversimplifying European history, for modern science and the modern commercial-urban-industrial social system grew up in medieval and early modern Europe, the political order of which was absolute monarchy ruling over an agrarian society with a social system of hereditary classes or castes. Therefore, political democracy was definitely not present when capitalism and science first arose.

On the other hand, no one states that modern science grew by itself, without correlation to its politico-social environment. In fact, it is undeniable that there are intimate correlations between the assertion of individualism (which, in European history, was manifested in nonconformist thinking on religion, unorthodox scholarship, and innovative artistic expressions), general material and technological progress (among others, printing), and the gradual recognition or premonition of the usefulness of sciences in promoting such progress (F. Bacon as the representative). These significant factors promoted the growth of capitalism, modern science, and new political elements such as the widening of religious toleration, the upgrading of the social status of intellectuals, the recognition of politics as a rational 'science' (Machiavelli as a representative), etc., which latter generally exerted positive influences on the former two aspects, particularly on the growth of science. In Europe, the idea of theoretical sciences had much earlier origins, both in Greek achievements and in Christian belief, than the above-mentioned premonition that it might serve material progress. However, only the combined humanist, religious, and political great leap forward that occurred consecutively and concurrently in Western Europe helped burgeoning modern science (not technology, which was relatively free to develop) to push over the weighty limitations of feudal mentality. In medieval Europe, the main inhibition of science, and modern rationalism in general, was also a territorially unified ideological rule, which was enforced by both lay and clerical authorities. But its oriental counterpart was different: first, there was no clerical authority in traditional China to act as a check; and secondly, European political authorities were fragmented and mutually restraining. These Western conditions, in contrast to China, provided independent, nonconformist thinkers with a substantial 'degree of freedom'.

When we compare Christianity and Confucianism more specifically, we take notice of the following major differences. As we have mentioned before, Confucianism was exclusively politico-ethical whereas Christianity did not discourage interest in God's cosmological design; besides, since Thomas Aquinas, Christian theology tried hard to use formal (Greek) logic. Although in Confucianism, especially after Mencius, some elements championed the rights of the populace, the explicit Confucian attitude, has always been understood (correctly) as that of an obsequious scholar uncritically serving the incumbent imperial polity. One convincing piece of evidence is that when the Taiping Rebellion started as a genuine mass movement, its leaders advocated anti-Confucian ideology; in fact, they declared that they were 'Christians'. On the other hand, both Catholic and Protestant churches imposed harsh limits on people's independent thinking and pursuits, especially on matters of God's cosmological design; but the actual historical development in Western Europe, its religious wars and later enlightenment, substantially and progressively eroded their authority to impose limits. While Europe boasted of its progress toward 'enlightenment', the last two Chinese monarchical dynasties, the Ming and Qing, were characterised by mounting rigidity in ideological control. That is why 'rational' and pragmatic China, with no religious taboos on a cosmological speculation, did not develop modern science whereas the world of Latin Christendom did. This fact seems to contradict my previously mentioned 'Spencer Gauge', but the contradiction is illusory: the foregoing examination of the Sino-Western contrast makes this clear. With regard to understanding nature, the essential differences between the two sides is almost that between indifference and overzealousness.

14 No Political Rationalism: Another Serious Defect

According to Max Weber, capitalism is, in the economic sphere, a ceaseless striving for greater and greater rationality. In the realm of politics, it is political rationalism, the rule of law, or to go one step further, constitutionalism. This poses a sharp contrast to Confucian 'benevolent government', which was described by Fairbank as a[51] 'highly personal administration of the government, following a "virtue ethic" which attached importance to the qualities of individual personality rather than a "command ethic" which laid emphasis upon an impersonal and higher law.' (See also note 14 of the Introduction.) According to Needham, since the transition from the School of Legalists to Confucianism at the very beginning of traditional China, 'during the period of transition from feudalism to bureaucratism . . . the Chinese early acquired a great distaste for precisely formulated abstracted codified law'.[52]

Chinese autocracy placed absolute and limitless power in the hands of the emperor. Yet, a few devoted and courageous ministers admonished their supreme authorities. One celebrated instance was Wei Zheng who said to Emperor Tang Tai-zong, among other things, 'Listen to both sides and you will be enlightened, heed only one side and you will be benighted.' But there could never appear in Chinese history an Edward Coke addressing his king, James I: 'The king should not be under any man, but under God and the law.' There has never been constitutionalism in China.

That the emperor's personal arbitration constituted the final say was philosophically consistent with the lack of objective criteria, which is a fatal weakness in developing a scientific way of thinking. In Essay One (Chapter 2) I argued that Chinese thinkers rarely had even a dim idea of scientific axiomatisation; in fact they seldom demonstrated a clear concept of seeking natural laws as inviolable statements of phenomenological regularities or as deterministic expressions between causes and effects. P. Young published a paper in 1979 that offered an opportunity for a confrontation of ideas on historical conceptions of time. He wrote,[53] 'The traditional Chinese conception of time was that events occurred in time in a cyclical manner, perhaps because of the strong agricultural aspect of the society.' To him, there was not only a cyclical view of time, but also a retrogressive attitude:[54]

> Thus the Chinese did in effect have a Garden of Eden to look back to . . . They looked backward to the ideal state which Confucius had spoken about . . . Hence, to attain the ideal they would have to copy what had been done in the past — this was past-aspiring compared to Western future-aspiring.

In the next volume of the same periodical Needham expressed his refutation, which had been published 13 years earlier.[55]

> That the Confucian outlook was essentially backward-looking is a thesis which in the light of the evidence brought forward in this essay alone cannot stand a closer look . . . [and] if their Chinese time had not been linear it is hardly conceivable that they would have worked with such historical-mindedness and such bee-like industry.

It seems that arguments supporting both positions on the concept of time in traditional China are available, whereas we notice that Lao Zi's well known maxim addresses itself directly to the relationship between natural laws and time:

> Dao ke dao, fei chang dao;
> Ming ke ming, fei chang ming.

The English translation of these twelve Han characters can be:[56]

> One may state a law,
> But it cannot be changeless;
> One may define a thing.
> But the definition is not constant.

Hasn't scholars' interest in looking for deterministic causalities been dampened by this pessimism (or prophecy)? (If Dong Zhong-shu represents orthodoxy, which asserts 'Heaven changeth not, likewise the Dao (Tao) changeth not', then here we have a Taoist unorthodoxy.)

15 Debates Around Zhu Xi's Historic Role

The historic role played by the Neo-Confucianism of the Cheng brothers (Chen Hao, 1032–85; Cheng Yi, 1033–77) and Zhu Xi (1130–1200) was again the focus of sharp debate. Needham asserts that Zhu was the first to have 'appreciated the true nature of fossils, 400 years before Leonardo da Vinci',[57] that he compared crystals such as flakes and salt, and made other observational and experimental studies.[58] Zhu has been described by him as 'both the Herbert Spencer and the Thomas Aquinas of China',[59] and 'somewhat reminiscent of Aristotle',[60] as a man 'whose philosophy is through and through naturalistic'.[61] On the other hand, as we have seen (note 15), Hartwell thought Zhu's orthodoxy marks the beginning of traditional China — doubtless, a retrogression.

Zhu Xi encouraged scholars to investigate things and to seek to reach the *li*. However, Sinologists differ in the understanding of the *li*. I accept Nakayama's view:[62] 'Within the Neo-Confucian concept of *li*, moral issues and the law of nature remained undifferent-tiated; thus it plays an inhibitory role in the development of the modern way of thinking.' Let us examine one of Zhu's best known statements:[63]

> In every human mind there is the knowing faculty; and in everything, there is its reason. The incompleteness of our knowledge is due to our insufficiency in investigating into the reason of things. The student must go to all things under heaven, beginning with the known principles and seeking to reach the utmost. After

sufficient labor has been devoted to it, the day will come when all things will suddenly become clear and intelligible.

So he did say that 'all things under heaven', not just human relationships, were at issue. According to modern scholars of Neo-Confucianism, his interest in nature was influenced by his adoption of some Buddhist and Taoist doctrines. Zhu departed from the traditional Chinese concept that human beings are harmoniously-fitting parts of nature and distinguished man from nature as having a knowing faculty. He, moreover, encouraged man to investigate the reason in everything.

If Zhu's disciples — and there were millions of them in subsequent centuries — followed instructions like these, could we expect that the 'reasons' would in due course have developed from empirical rules and phenomenological regularities into theoretical laws and general principles? If the ancestors of the Chinese people had really done this, developed and accumulated their positive results, then China could no doubt have been another, even the first, cradle of modern science. But this was not the actual story. It was as if Zhu's Neo-Confucianism ruled China from 1313 to 1905, while Zhu's teaching about nature did not function in its proper way. But I think they did function in their proper way; the problem is to understand their functioning properly.

In Zhu's above statement the whole procedure of investigating things until 'the utmost principles' were reached was, or had to be understood as, subjective activities of the scholar. Originating with Confucius and his proximate followers, and becoming a kind of 'religious' practice in China long before Zhu Xi, scholars were required to 'cultivate character' and to 'refine (or mellow) personality'; the goal was explicitly subjective and personal. In Zhu Xi's 'all things clear and intelligible' and the unattainable 'utmost', no objective standard was implied. From the wording 'sufficient labor' and 'suddenly become clear', it is obvious that the truthfulness of knowledge was to be judged by individual, subjective criteria. Just like 'the Tao eternal, the Tao beyond speech, that cannot be spoken about', Zhu's *li* was presumed to be transcendent, but it was also functionally non-transferable. With 'investigating all things' it seems that Zhu Xi deviated from the traditional Chinese subjectivism; but in fact he followed this heritage. It was at best a new dimension added to the traditional conception: one should strive to attain the mental attitude with which one could confront every problem of human affairs or of nature, with equanimity and calmness. This approach represents the time-honoured Chinese ataraxia (common to Confucianism, Taoism, and Buddhism). With this mentality one 'understands' everything 'intelligibly'. In traditional phraseology, this state of mind is qing-jing wu-wei, the doctrine of ataraxy and non-interference.

Nakayama wrote:[64]

> Needham claims that the Neo-Confucian term *li* is a basic concept of natural organization, features of which were transmitted by Leibniz to the West. There they played a clearly definable part in the formation of modern science.

Here we have two Needhamite hypotheses. One is about Chinese thought influencing Leibniz; the other is about the relation of Chinese thought and Leibnizan philosophy to science. But we have to be careful about which science in this context Needham refers to: modern or future? Nakayama conveys correctly the first Needhamite hypothesis, but the second part needs clarification on two counts. First, for the 'formation of modern science' Leibniz (1646–1716) was somewhat late; Descartes could fill this place much better, not only chronologically but also philosophically — particularly if another (though closely related, but in addition to these two) Needhamite hypothesis, that Leibniz did represent the so-called organicist philosophy, is true. Secondly, Needham thinks that the 'Chinese correlativism', which influenced Leibniz profoundly, will play a major (if undefined) part in guiding the *future* of science (see Essay Three, Chapter 4).

16 Tentative Resolutions of Two Paradoxes

Now the proposed theory of this work itself is confronted with a historical paradox: if the monopolistic Confucian rule was truly the nemesis of medieval China's progress, then how can we explain the appearance in Chinese history of the so-called Medieval Flowering between the eighth and thirteenth centuries? If my basic hypothesis is correct, there must have been relaxation of ideological control during the Flowering. It seems that this inference is correct. At the outset of the Medieval Flowering, the centre of Chinese culture moved southward from the Yellow River valley to the Yangtze valley. One could assume the limit of strict control at a time of massive migration, nomadic invasion, warlord divisionism, and the comparatively peaceful admixture of foreign cultures, most prominently Buddhism and Islam.

According to Fan Wen-lan,[65] the time from the third to the sixth centuries saw enormous economic and cultural developments in the Yangtze valley. The Sui-Tang unification carried over this spirit of toleration and pluralism. In fact, the founders of both dynasties had non-Han origins. Indeed, we have some notable signs to bolster the assumption that Confucian orthodoxy was not rigidly enforced during this period. The middle of the Tang witnessed the ascendancy of an empress with patent non-Confucian personal characteristics. Hartwell pointed out that the novel emphasis on the historical analogism in the so-called 'new history' of the Tang and Song periods represented a radical reorientation from the moral didactic orthodoxy, and that the new scholarship received its initial impetus during the intellectual ferment of the half century following the rebellion of An Lu-shan.[66] To summarise the situation, we can quote B. Schwartz:[67]

> We know that during the centuries which stretched from the end of the Latter Han dynasty to the middle of the Tang dynasty Confucianism ceased to exercise a dominant hold even on those strata whose class interests it presumably represents.

In all probability my readers would raise another paradoxical comparison. That is with Japan, China's neighbouring nation, apparently bearing a number of similarities, enigmatically including the Confucian tradition. Again I will mention a couple of important historical developments and features as signs that show the essential differences between the two nations.

For centuries a classic difference has been on the one hand the blind complacency of the 'Middle Kingdom' and on the other the willingness of the 'Land Where the Sun Rises' to borrow foreign culture. Moreover, as mentioned in the Introduction, prolonged feudal separatism made Japan, like medieval Europe, a political environment totally unlike imperially unified China. The Tokugawa period (1603–1868) had some additional features that served as preparatory conditions for the subsequent rapid modernisation under the Meiji reign. Compulsory urbanisation in each of the 260-odd domains, produced a vigorous development of financial economy and a higher average literacy than in China. Owing to the local schools for commoners, by the time of the Restoration, more than 40 per cent of all Japanese boys and about 10 per cent of girls had formal education outside of their homes.[68] Because of 'stratified' and localised feudal loyalties, there was also a high average level of political consciousness among Japanese samurai and commoner elite.[69] Feudal separatism and competition sustained and encouraged an intellectual eclectism,[70] one component of which, the enthusiasm for 'Dutch learning', turned out to be a vital national undertaking. For a few decades before 1868, the year of the Meiji Restoration, powerful daimyo competed or co-operated in matters of rudimentary industrialisation, military build-up, educational efforts, and foreign affairs.[71] When the revolution did happen, this eclectic attitude was embodied in an almost conditionless willingness to try things new and to follow what was better. As a matter of fact, the Chinese Tong-zhi Restoration and the subsequent Westernisation Movement happened contemporaneously with the Japanese agitation. Yet the results on the two sides were so different. With one of the theoretical assumptions of this paper, that is the theorem that 'software decides', it is not hard to understand the contrast. As early as at the end of the Tokugawa Shogunate, the idea of 'government through public discussion' was developed. After the Restoration, this idea was replaced by 'respect for public opinion', and in the heat of the civil wars of 1868, the young emperor declared: 'Knowledge shall be sought throughout the world, so as to strengthen the foundation of imperial rule.' In other words, Japan at this time became fully aware of the necessity of changing its 'substance', (see note 16 and Section 15 of Essay One, Chapter 2), whereas the Manchu regime in China held on to its 'substance' as its lifeblood.

To pick up just one piece of indicative evidence: in 1872 Fukuzawa published the first article of his 18-instalment 'Encouragement of Learning', which quickly sold 200,000 copies — just four years after the Restoration! The gist of the article is that the hierarchical division of social status will soon be abolished; from now on a man's worth depends on his useful contributions to society, which in turn is decided by his learning.[72]

At the same time, China presents quite a different picture, with dynastic despotism, bureaucratic corruption, Confucian civil service examinations, subordination to and exploitation by imperialist powers, and general benightedness of its people. In retrospect, the Boxer Rebellion on the Chinese side contrasts conspicuously with the diplomatic efforts to revise unequal treaties on the Japanese side. Small wonder that the ensuing developments on the two sides have been so diametrically different. No difference is more potent or generative, or better illustrates my theme, than the fact that not long after the Meiji Restoration, Japan succeeded in enacting a constitution, and thus laid the foundation of its modern politics, while after more than 100 years China is today still struggling to produce such a document!

17 The Principle of Intellectual Activation and Creativity

In Section 12, 'Some Illustrations of Sino-Western Contrasts', we came across the large difference in the degree, or level, of social activation at the two ends of Eurasia. With regard to the growth of modernity, and modern science in particular, the common and necessary ingredient of various types of social activity was intellectuality and creativity. Therefore we can formulate a general principle, or criterion, to summarise the factors that are responsible for the development or non-development of modernity.

Modern science requires the growth of 'modernity' in general as a prerequisite.

The growth of modernity, and therefore the growth of science in particular, depends on the degree (within a sizeable realm of common culture) of intellectual activation and creativity, which have extensive, intensive, pluralistic and inheritable aspects, and which serve constructive, innovative, explanatory, and exploratory purposes.

This formulation I call 'the principle of the degree of intellectual activation and creativity'.

Because of the following facts, traditional China was characteristically low in the degree of intellectual activation and creativity:

(a) The nature of the politico-social structure — an imperially unified, thoroughly-indoctrinated officialdom ruling over a vast territory and agricultural population.

(b) An autocratic, introspective, and immobile ruling ideology. It helped to consolidate Chinese autocracy; it was solely interested in politico-ethical relationships; it advocated a changeless world order; it muffled enterprising and independent spirits which are necessary in pursuing exact knowledge; and it lacked objectivity as an epistemological criterion.

(c) An effective bureaucratic straitjacket. In traditional China the time-honoured, 'democratic' system of civil service examinations was the linchpin between political control and ideological indoctrination. T. Najita, a scholar of Japan, has argued that the leaders of Tokugawa Japan were not orthodox but ideologically eclectic, and that they had no bureaucratic apparatus to enforce orthodoxy.[73] It is safe to say that this Japanese situation was just the opposite of its Chinese counterpart. In particular, the Chinese examination system was the bureaucratic apparatus that effectively sustained an orthodox ideology in China for centuries.[74]

In the Introduction we have mentioned that Hu Shi thought that since the Tang there had already been five recognisable Chinese 'renaissances'. But Liang Qi-chao (1873–1929), a great man of letters and reformer, had a better summary of the history of Chinese thought:[75]

> In our country since the Qin (255–207 BC), the movements that have truly developed into 'periodic tides of thought' were the Classicism of the Han (206 BC–220 AD), the Buddhism of the Sui (590–617) and Tang (618–906), the Neo-Confucianism of the Song (920–1279) and Ming (1368–1643), and the School of Empirical Research of the Qing (1644–1911) — only these four.

'Only these four', — indeed, I think we are justified in deriving a salient feature of the dominant Chinese systems of thought (as I have already mentioned in the Introduction) by showing that these 'tides of thought' were essentially very much alike. This statement includes the last 'tide' because of its contents, not its methodology. Therefore, with a minor exception, none of the 'tides' shared any common interest with the intellectual ingredients that comprised modern science. To put it figuratively, the dominant Chinese thoughts offered too narrow and too exclusive a soil for the tree of modern science to grow. I agree with Needham that the Chinese system was integrative and absorptive *vis-à-vis* surrounding ethnic groups (as a one-way traffic). But it was also extremely unintegrative and un-absorptive *vis-à-vis* more distant systems of culture and thought. In other words, during the long course of traditional China, the Han people 'Hanised' the surrounding peoples to a great extent; but, on the one hand, it is impossible to 'Hanise' the West, and on the other hand, it is a very hard task to Westernise China.

18 A Brief Summary of the Arguments Presented

The Needham Puzzle, as an inevitable sequel of voluminous research, has been an interesting and important question.

On the other hand, by juxtaposing the presupposition (China had marvellous scientific and technological past) which backs the Needham Puzzle and the dichotomous character of China's history (instability within continuity), or simply China's apparent inertia, we arrive at a paradox (the alleged nearness to generate rational science in medieval China presents an incongruous picture to the salient difficulty of implanting science in modern China) that resolves (or invalidates) the Puzzle itself.

To re-state, our paradox is: China's glorious inventive past contrasts with China's persistent resistance, past and present, to 'modernisation'.

Therefore, although the Feng Puzzle bears a formal similarity to the Needham Puzzle, logically it is very different, because the two were based on very different premises: Needham's was 'medieval China was so incredibly marvellous' and Feng's was '20th century China still shows obvious incompetence'.

Modern science grew as the result of a conscious and broadly based campaign among intellectual elites, and through successive generations. It required persistent efforts in rational inquiries about nature — which have been progressively characterised by logical clarity, experimental verifiability, and mathematical accuracy — to lay the theoretical and methodological infrastructure of modern science. China, transitional as well as traditional — where a unified combination of autocratic rule and ideological control has never been dismantled, cracked, or even relaxed — cannot initiate such a campaign and cannot sustain such efforts.

The degree of intellectual activation and creativity summarises the factors that are responsible for the development or non-development of modernity, and modern science in particular.

It should serve us as a useful criterion to judge advantageous and disadvantageous conditions in terms of a society's potential towards economic, political, cultural, and scientific progress.

Because I have, in a sense ventured 'axiomatisation' in historical inquiry, (cf. 'Explanations and Acknowledgements', p. 4), I would like to share with you this amusing — maybe also profound — story:

Once Confucius and his disciples found themselves in a very difficult situation, and he noticed Zi-gong, one of his students, was displeased. Then he asked Zi-gong:

'Do you think me a learned, well-read man?'

'Certainly,' replied Zi-gong, 'Aren't you?'

'Not at all,' said Confucius. 'I have simply grasped one thread which links up the rest.'

— Recounted in Sima Qian (145-*c.* 89 BC), 'Confucius' in *Shi Ji*

Notes

1. J. Needham, *Science and Civilization in China*, vol. I (Cambridge University Press, 1954), p. 119.
2. J. Needham, *The Grand Titration* (University of Toronto Press, 1969); p. 11. The title of this book will be abbreviated as TGT below.
3. TGT, p. 154.
4. Ibid.
5. In his *Science in Traditional China* (The Chinese University Press, Hong Kong, 1981) Needham tells us (pp. 2–3) that this question, 'why modern science originated only in Europe', was first mentioned to him in the early 1930s by several young Chinese research workers who went to Cambridge for their doctorates.

It was not until January 1983, when I was working specifically on a Sino-Japanese comparison, that I learned about Boulding's paper (K. Boulding, 'The Great Laws of Change' in A. Tang, F. Westfield, and J. Worley (eds), *Evolution, Welfare, and Time in Economics* (Lexington Books, 1976)) and Tang's paper (A. Tang, 'China's Agricultural Legacy', *Economic Development and Cultural Change*, vol. 28, no. 1 (1979), pp. 1–22). These two papers both discussed the Needham Puzzle, naming it the Needham problem or the Needham paradox. Understandably I was much delighted to read ideas which are similar to what I have arrived at independently:

> In China, however, the resistance to new basic ideas was very well organized because of the internal unity and homogeneity of the society and the existence of a self-perpetuating Mandarin class. In Europe, the resistances were more scattered and fragmented. In China the church and state were virtually a single organization, at least in the Confucian elements in society, whereas in Europe there had always been separation between the emperor and the pope, not to

> mention the separation within the church, in the Protestant Reformation, and the division of Europe into many different states. (Boulding, pp. 9–10)

Although Needham likes to narrate his personal story of working on Chinese civilisation, in his recent book quoted above, *Science in Traditional China*, what amazed me is a related passage in which Needham frankly revealed the transition in his view toward the Chinese (p. 3).

> because I found that the more I got to know these friends from China the more exactly like my own their minds seemed to be, certainly in their intellectual capacity; and this raised in an acute form, therefore, the question of why modern science, the 'new or experimental' philospphy of the time of Galileo, had arisen only in European culture and not in China or India.

(So, here we have a theorem that says 'brain decides'.) In the light of this confession, I found myself better able to understand Needham's stand, which he defended more than once (p. ix):

> if sometimes we have written like barristers pleading a case, or sometimes over-emphasize the Chinese contributions, it has been consciously to redress a balance, which in the past tilted over much too far on the other side. We were out to redress a secular injustice and misunderstanding.

I think it will not take long for people to see whether this strategy is truly useful for either a Western or a Chinese audience.

6. A. C. Graham, 'China, Europe, and the Origin of Modern Science: Needham's *The Grand Titration*' in Nakayama & Sivin (eds), *Chinese Science* (The MIT Press, Cambridge, Mass., 1973), p. 45. (Hereafter, *Chinese Science*.)

7. Ibid., pp. 45–6.

8. Ibid., p. 51.

9. Since my manuscript has been criticised on this and a few related points (although to me, to my advisors, and to most of my audience, they are crystal clear and correct), let me add an explanatory remark.

Formally (logically), a positive question and a corresponding negative question do not form 'the two sides of a coin', because while the positive question is one side, the negative questions are many-sided, or each one of them is one among a list. For instance, why gold is gold is a positive question, whereas its corresponding negative questions are why gold is not white, not black, not red . . . or, why silver is not gold, why iron is not gold, why mercury is not gold . . . On the question of the emergence of modern science, formally the positive question and the negative questions conform to this relationship, particularly the latter case. (Even so, raising some other negative questions is not pointless: the Indians, the Greeks, the Arabians, the Africans . . . all are entitled to ask the negative questions in relation to their own history, that is to say, to write and to argue about their nations' 'great inertia'. Even physical questions such as why silver is not gold, etc. cannot be dismissed *a priori* as pointless. Thermodynamics was established by examining various processes which are not ruled out by mechanical laws. Historically, it was because of asking and answering why it is not possible to put various *perpetuum mobile* designs into practice and why it is not possible to raise the efficiency of steam engines by leaps and bounds. Then thermodynamics emerged, of which the three basic laws all took the form: 'It is impossible to realize a process that . . .' And as we know, they turned out to be very *positive* and productive statements.) Thanks to improved knowledge of the Oriental scientific past in the last several decades, medieval China has been singled out as a plausible candidate for generating modern science. Then we have a two-sided coin. Graham's logical inconsistency came from switching unconsciously from particularity to generality.

10. TGT, p. 213.

11. Ibid.

12. Ibid., p. 284.

13. There were, in fact, tenuous relations between early technology and science (see note 23). The earlier, the more tenuous the relations were. Correspondingly, and evidently, there has been a growing intimacy between science and technology in modern times.

14. For China, the view of 'implanting modern science into . . .' logically contradicts Needham's ideas. In 1975 Needham remarked, 'The idea, for example, that Chinese scientific development has been "just a transplant", something out of keeping with the cultural inheritance, is sheer nonsense . . . In the light of all this, some estimates of China with relation to science need complete revision, in fact, relegation to the scrap-heap.' (J. Needham, 'History and Human Value: A Chinese Perspective for World Science and Technology', 1975, Occasional Paper, McGill University, p. 8.)

When I first perceived in the spring of 1981 the basic Chinese paradox — that despite its glorious medieval progress, the nation today still shows such stubborn ineptitude at modernisation — I did not know that I was also trying to rise up from Needham's 'scrap heap'! By linking autocratic tradition to totalitarian reality and pointing out the parallelism in ideological control, I will certainly offend one side of the Taiwan Straits. By accusing 'Confucianism' as the ideological source of the great inertia I will equally offend the other side of the Straits; while the 'third China', by which I mean Hong Kong, is apparently overcrowded. It seems no China will accept me yet my family is still there! Once freed from political suppression and endowed with the freedom to study, to think and to discuss, I simply cannot help speaking out. My personal sacrifice does not match my nation's long suffering and my compatriots' common agony. Day in and day out, my embittered heart urges me to contribute more directly (in comparison with doing physics) to the rescue of my poor old China — at this time, and by my nature, through honest and assiduous scholarly pursuits.

15. R. Hartwell, 'Historical Analogism, Public Policy, and Social Science in Eleventh- and Twelfth-Century China', *The American Historical Review*, vol. 76, no. 3 (June 1971), p. 727.

16. For instance, Cong Han-xiang, 'Zhog-guo Feng-jian She-hui Nei Zi-beng Zhu-yi Men-ya Zhu Wen-ti (Several Questions Concerning the Emerging Capitalism in Feudal China)', *Li-shi Yan-jiu (Historical Research)*, no. 84 (Beijing, Dec. 1963), pp. 19–64.

Several pages in Edwin Reischauer and John Fairbank, *East Asia: The Great Tradition* (Houghton Mifflin Company, Boston, 1960), pp. 333–7, also provide a useful reference in English.

17. Professor Robert Winter (1886–) of Beijing University said to me in January 1980: 'I love China, a country of philosophers — even if they are in rags.' He meant China's ancient cultural achievements — I am sure he did not mean Maoist China.

18. Einstein's letter to J. E. Sweitzer, quoted in *Chinese Science*, p. 50.

19. In 1978 I finished an informational (that is, mathematical) study of the readability of Han character writings, and found the equal spacing between every two consecutive square characters, or the non-clustering of characters into *words*, causes the largest loss of useful information. (Consequently, the intentional omission of punctuation, which was a common practice in traditional China, and the lack of capitals in Han characters, added more difficulty to readability.) One of my delights in the past few years is that this theory of mine has been fully

demonstrated in my experience of helping some American friends to read Chinese newspapers. The work is included in the fifth chapter of my book, *Theoretical Physics for Engineers*, which was printed and distributed by the printing office of Zhejiang University, Hangzhou, China.

20. TGT, pp. 152–3.

21. Ibid., p. 150.

22. C. Totman, *Japan before Perry* (University of California Press, 1981), p. 189.

23. N. Sivin, 'Science in China's Past' in Leo A. Orleans (ed.), *Science in Contemporary China* (Stanford University Press, 1980), pp. 4–5.

But the mid-nineteenth century is a bit late for the interdependence of science and technology. Considering the works of Sadi Carnot, Michael Faraday (and Humphrey Davy before him), Hustus von Liebig, etc., the time should at least be shifted two to three decades earlier, that is to say the first half instead of the middle of the nineteenth century.

Sivin's clarification of this point is particularly interesting to me because it drives home the shallowness of the question: Those people had so many technological inventions, why did they not 'invent' modern science? From now on we shall perhaps no longer ask the question this way, because these two aspects are recognised as not directly related in history. In fact, the book's editor, Orleans, also expressed a very interesting thought when he raised the question: 'Can China achieve its goal of "four modernization" without also introducing the "fifth modernization" — a liberalization of the system?' (p. xxxii). To me, Sivin's history (the separation of the early science and technology, perhaps with the exception of medicine) and Orleans' reality (the likely decisive role of politics) are not only mutually complementary but unified. I was much delighted to read their book a year after I had stated my 'theorem'; both history and reality demonstrate that 'software' (political institutions, dominant ideologies) is more important than 'hardware' (the level of economic development, technologies, hard sciences). Here is a brief explanation of this 'theorem':

Needham and others revealed that traditional China had numerous technological achievements, yet there could not be a 'modernity revolution' there and then (hence no emergence of modern science). The major reason was that traditional China's territorially unified autocracy, combined with an equally unified ideological control, was too harsh. 'Software' decides.

Contemporary China has been busy about 'hardware' modernisation for more than three decades, yet in various fields there have been very little, or no, or negative successes — because there is no 'software' modernisation. (The people are allowed no say at all on vital questions of the nation as well as of their own lives. As a result, the dictators' obviously foolish and cruel whimsies ruined the nation's economy, culture, morale, and moralities.)

24. Namely Yu Lan Fung, eminent professor of philosophy, Beijing University, China. The spelling of the name is updated according to the pin-yin system.

25. Feng You-lan, 'Why China Has No Science', note 15 of the Introduction. Henceforth, Feng, 'Why'.

26. S. Bennet, 'Chinese Science: Theory and Practice', *Philosophy East and West*, vol. 28 (1979), p. 439.

27. Feng, 'Why', p. 237.

28. This was not my original perception of the question. When I read Feng's 'Why' I had already several times redrafted my own paper.

29. For example, H. H. Gerth & C. W. Mills, *From Max Weber, Essays in Sociology* (Oxford University Press, Oxford, 1970–72 reprint), p. 431.

30. Feng, 'Why', p. 237.

31. Ibid.

32. Ibid., p. 238.
33. Ibid., p. 259.
34. Ibid., p. 261.
35. Sivin (ed.), *Science and Technology in East Asia* (Science History Publications, New York, 1977), p. xx.
36. Sivin, 'Science in China's Past', note 23 above, p. 5.
37. Most of the Confucian quotations and basic ideas are not given references, since they are very general ones. In China, particularly after the political campaign named 'Criticizing Lin Biao and Denouncing Confucius' around 1974, people became very familiar with these quotations from numerous official pamphlets, many English versions of which can be found in the *Beijing Review*. It is obvious that the theme of this paper has nothing in common with 'Criticizing Lin Biao and Denouncing Confucius'; it takes a diametrically opposed stand against the spirit of that government-controlled campaign.
38. J. Spence, 'Imperial China: A Chinese View' in C. Worwich & J. Spence, *Imperial China: Photographs, 1850–1912* (Pennwick Publishing Inc., 1978), p. 9.
39. J. K. Fairbank, *The US and China,* 4th edn (Harvard University Press, 1979), p. 55. (Henceforth, *US & China.*)
40. *US & China*, p. 47.
41. K. Yabuuti, 'Chinese Astronomy: Development and Limiting Factors' in *Chinese Science*, pp. 93–4.
42. Needham, 'History and Human Value', note 14 above, p. 4.
43. Yabuuti, note 41 above, p. 95.
44. Ibid., pp. 97–8.
45. Ssu-Yu Teng and J. K. Fairbank, *China's Response to the West* (Atheneum, 1968), p. 19.
46. Twitchett's statement that during the Southern (Later) Song, 'for the first time in Chinese history, to build and maintain a great navy', may not be accurate. According to Fan Wen-lan, the State of Wu, during the Three Kingdom period (about nine centuries earlier than the Southern Song), had a navy of more than 10,000 men. See D. C. Twitchett, 'The Middle Kingdom' in A. Toynbee (ed.), *Half the World* (Thames and Hudson, London, 1973), p. 70; and Fan Wen-lan, *Zhongguo Tong-shi Jian-bian (An Outlined Chinese History*), vol. I (Ren-min Chu-ban-she, Beijing, 1955, 3rd edn, 1961, 7th reprint), p. 7.

In view of these facts and the early Ming termination of naval power, we see here a Chinese 'naval inertia'.

47. Ray Huang, *1587, A Year of No Significance: the Ming Dynasty in Decline* (Yale University Press, 1981), p. 160.
48. D. C. Twitchett, note 46 above, p. 71.
49. *Encyclopaedia Britannica*, 11th edn (1911), vol. XXVII: p. 752.
50. TGT, p. 138.
51. *US & China*, p. 45.
52. TGT, p. 36. See also note 29 of Essay One (Chapter 2).
53. P. Young, 'The Sociology of Time: Histories and Historians in the Cultures of the West and China', *Leonardo*, vol. 9, no. 3 (1976), p. 205.
54. Ibid., p. 207.
55. Needham, 'Time and History in China and the West', *Leonardo*, vol. 13, no. 10 (Oct. 1977), p. 235.
56. In for instance T. D. Lee, 'Elementary Particles', *Physics Today*, vol. 13, no. 10 (Oct. 1960), p. 36, the wording of the translation is somewhat different:

   The Tao that can be stated
   cannot be the absolute Tao
   The name that can be given
   cannot be the permanent name.

57. TGT, p. 149.
58. Ibid., p. 33.
59. Needham, *Within the Four Seas* (George Allen & Unwin, London, 1969), p. 66.
60. Ibid., p. 94.
61. TGT, p. 149.
62. S. Nakayama, 'Joseph Needham, Organic Philosopher' in *Chinese Science*, p. 40.
63. For example, L. Carrington Goodrich, *A Short History of the Chinese People* (Harper and Brothers Publishers, New York, 1943), p. 158.
64. Nakayama, note 62 above, p. 40.
65. Fan Wen-lan, *Zhong-guo Tong-shi Jian-bian (An Outlined Chinese History)*, vol. II (Ren-min Chu-ban-she, Beijing, 1957), chapters 5 & 6.
66. R. Hartwell, note 15 above, p. 694.
67. B. Schwartz, Foreword to Liang Ch'i-chao (I. C. Y. Hsu, translator), *Intellectual Trends in the Ch'ing Period* (Harvard University Press, 1959), pp. xi–xii.
68. R. Dore, *Education in Tokugawa Japan* (Routledge and Kegan Paul Ltd, London, 1965), p. 254 and Appendix I.
69. For the political role of the commoner elite and the social strength of Tokugawa Japan, see for instance C. Totman, *Japan before Perry* (University of California Press, 1981), p. 135.
70. About eclecticism on political thinking, see Totman, ibid., p. 178. See also note 73 below.
71. See, for example, Thomas Smith, *Political Change and Industrial Development in Japan* (Stanford University Press, 1955), Chapter 1.
72. Y. Fukuzawa, *The Autobiography of Yukichi Fukuzawa* (Columbia University Press, 1968), pp. 390–7.
73. T. Najita, *The Intellectual Foundations of Modern Japanese Politics* (The University of Chicago Press, 1974), p. 31.
74. In B. Russell's *The Problem of China* (note 12 of the Introduction), pp. 44–7, there are analyses and denunciations of this harmful system, both by Russell himself and a Chinese historian, Li Ung Bing.
75. Liang Ch'i Ch'ao, note 67 above, p. 19.

4 SCIENTIFIC PHILOSOPHIES: CHINA'S PAST – THE WORLD'S FUTURE?

1 'Joseph Needham, Organic Philosopher'

Needham made great contributions in summarising the Chinese history of science and technology. His pioneering work greatly enhanced interest in the scientific past of eastern Asia. I believe that we can also see in his comparative perspectives a number of questions which are significant in understanding the growth of modern science. Here I will discuss some Needhamite speculations from a critical stand. In view of my analysis, the full scope of which is more comprehensive than this chapter covers, I believe that Needham must be appreciated as a remarkable encyclopaedist, but not as a great philosophical historian; someone who raised many more important questions than he could satisfactorily answer.

Needham was a biochemist, embryologist, and also philosopher and historian of science until 1937 when he was 37. Since that year the history of Chinese science fully occupied his attention. These experiences, in addition to his religiosity, I believe, formed the background from which he has been called by some historians as an 'organic philosopher'. According to S. Nakayama, in the 1920s 'Needham saw in Whiteheadian organismic philosophy the synthesis of mechanism and vitalism.'[1] In the four basic ingredients of his thought, again according to Nakayama,[2] the first was a strong antipathy to the mechanistic view that characterises many physical scientists', and the fourth, an 'inclination to synthesis over analysis'. And he also 'has an intrinsic propensity towards organism'.[3]

2 Needham's Faith in Chinese Correlativism

Needham saw in the history of Chinese science a great heritage of correlativism or organicism. He made an audacious hypothesis about the future development of science, which, as we can see from the second volume of his *Science and Civilization in China* and other writings, he stated again and again. Here is one passage from him:[4]

> An unexpected vista thus opens before our eyes — the possibility that while the philosophy of fortuitous concourses of atoms, stemming from the society of European mercantile city-states was essential for the construction of modern science in its 19th century form; the philosophy of organism, essential for the construction of modern science in its present and coming form, stemmed from the bureaucratic society of ancient and medieval China . . . All that our conclusion need be is that Chinese bureaucratism and the organicism which sprang from it may turn out to have been as necessary an element in the formation of the perfected worldview of science, as Greek mercantilism and the atomism to which it gave birth.

First let us clarify a question in the first half of the quote, although it is not directly related to the correlativism problem. What does it mean by 'the philosophy of fortuitous concourses of atoms'? Was this philosophy 'essential for the construction of modern science'? As I understand, historians generally agree that physics led modern science, while mechanics led physics. But mechanics evidently does not embody a 'philosophy of fortuitous concourses of atoms'. Rather, from Archimedes to Kepler and Galileo, and to Newton, that is, from the lever and a floating (or immersed) body to a falling particle or a revolving planet, and to a synthetic system of three neat laws plus universal gravity, mechanics represents the philosophy of the *simplest and most predictable interacting* system — in sharp contrast to *fortuitous concourses* of atoms.

Let us now come back to the problem of the organicist philosophy and our future science. With this regard, another Needham passage reads:[5]

> I would suggest that both this and the more extravagant extensions of the symbolic correlations of the five elements were exaggerations of certain basic ideas as valid in their way, and as valuable for the future history of human thought . . .

And a third passage reads:[6]

> The gigantic historical paradox remains that although Chinese civilization could not spontaneously produce 'modern' natural

> science, natural science would not perfect itself without the characteristic philosophy of Chinese civilization.

In other words, Chinese civilisation, which could not produce modern natural science, will be necessary for a 'futuristic' science. But can't we also understand this 'gigantic historical paradox' less sanguinely? In fact, I am not the first person to doubt whether there is any seed of truth whatever in this paradox. Nakayama wrote:[7]

> However, it is very doubtful that 'organism' in its Chinese version could ever take the role of a promoter of modern science. Some of the characteristics of organism that Needham considered uniquely Chinese might be found in other premodern cultures too.

According to Nakayama's observation, with which I agree, the so-called Chinese organism, or correlativism, is more or less a universally shared body of primitive or proto-philosophical thinking. It had not produced modern science in the past; and it is very unlikely to act as a booster of science today. But Needham looks at the situation very differently. In fact the implication of his logic is that thanks to the lack of development, some ancient cluster of proto-scientific or proto-philosophic ideas in China turns out to be 'necessary', 'essential', and 'valuable' for the future of science. That is to say, primitiveness is claimed to be a blessing. Ancient philosophy here becomes a kind of antique — as time elapses its value rises. Or it may be likened to a mineral deposit. The ancestors of the Chinese people did not have the perception and know-how to tap and utilise it; now a prophet tells us that it is going to benefit the whole world by illuminating the present and coming generations. It is indeed a gigantic historical irony.

3 What is Needham's Chinese Correlativism?

As indicated in the first of the above quotations, Needham believes the politico-social soil of Chinese correlativism is the bureaucratism in traditional China. He explains his Chinese correlativism this way:[8]

> In coordinative thinking, conceptions are not subsumed under one another, but placed side by side in a pattern, and things influence one another not by acts of mechanical causation, but by a kind of 'inductance'.

Another passage says:[9]

> The key word in Chinese thought is Order and above all Pattern (and, if I may whisper it for the first time, Organism) . . . Things behaved in particular ways not necessarily because of prior actions or impulsions of other things, but because their position in the ever-moving cyclical universe was such that they were endowed with intrinsic natures which made that behavior inevitable for them . . . They were thus parts in existential dependence upon the whole world-organism. And they reacted upon one another not so much by mechanical impulsion or causation as by a kind of mysterious resonance.

That is correct: there was plenty of mystery in ancient proto-science. Yet the spirit of modern physics, and modern science in general, is the gradual reduction of mystery. For many who regard this as the greatest merit of modern science, the statement that mysterious proto-science will turn out to provide the necessary impetus for forthcoming science really sounds extravagantly paradoxical. Unfortunately for modern mystics, one of the great lessons that we can derive from the history of science is that, anything that resolves mysteriousness is scientific, and that anything that retains or magnifies it is unscientific. On the other hand, we are aware that scientific mysteries, or wonderments, are endless. Great minds have been consistently fascinated and puzzled by wonderments, which, I think, may be more exactly named 'frontier wonderments' of a certain age and in a certain field of studies. As science and scholarship in general make progress, old wonderments are resolved and new ones appear, and the frontier is pushed ahead. That is what I mean by the resolution of mysteriousness by science. (Both mystery and wonderment pose questions. To me they differ in the degree of rationalism in the questioner's approach and articulation. Therefore, science stands further away from mystery but borders upon wonderment, although the latter is not a common word.) Before the inception of field physics, inductance, resonance, and 'actions at a distance' were mysterious. In order to resolve these mysteries,

M. Faraday and W. Thomson (Kelvin) introduced the concept of fields, the mathematical generalisation and experimental verification of which were successfully followed up by Maxwell, Hertz, and others.

4 What is Needham's Scientific Basis of His Methodological Revolution?

Needham tells us that since the time of 'Dalton, Huxley, and the mechanical materialists', science[10]

> has been obliged to become still more 'modern', to assimilate field physics . . . Deepening knowledge of biological phenomena, too, had necessitated a reformulation of scientific concepts in which the philosophy of organism has had a vital part to play.

In another passage Needham tells us that there is a great movement under way which rectifies the mechanistic ideas. For scientific supports, he again appeals to the development of field physics and biological sciences:[11]

> Here it is not possible to do more than mention the great movement of our time towards a rectification of the mechanical Newtonian universe by a better understanding of the meaning of natural organization . . . in its various ways, with varying acceptability of statement, it runs through all modern investigations in the methodology and the world-picture of the natural sciences — the numerous and remarkable development of field physics, the biological formulations which have put an end to the sterile strife between mechanism and vitalism . . .

Why does Needham think field physics support his philosophy of organism? From what we have quoted before we know that Needham thinks 'mechanical causation', or 'mechanical impulsion', is an outmoded concept for fundamental scientific causality. He makes a sharp distinction between 'billiard-ball' causality and 'field of force' causality:[12]

> On one world view, if a particle of matter occupied a particular

> point in space-time, it was because another particle had pushed it there; on the other, it was because it was taking up its place in a field of force alongside other particles similarly responsive. Causation was thus not 'particulate' but 'circumambient'.

Why does Needham think life science supports his philosophy of organism? Because he believes:[13]

> . . . the characteristic Chinese conception of causality in the world of Nature was something like that which the comparative physiologist has to form when he studies the nerve-net of coelenterates, or what has been called the 'endocrine orchestra' of mammals . . . it is now becoming probable that the higher nerve centres of mammals and man himself constitute a kind of reticulate continuum or 'nerve-net' much more flexible in nature than the traditional conceptions of telephone wires and exchange visualized . . . All this is quite a different mode of thought from the simpler 'particulate' or 'billiard-ball' view of causality, in which the prior impact of one thing is the sole cause of the motion of another.

How does Needham relate his belief in his evolution, or revolution, of scientific methodology to his scholarship in Chinese? Needham's understanding of the famous yet ill-defined 'Tao' could give us some hint of the working of his mind. In one context he tells us: 'In a way, the whole idea of the Tao was the idea of field of force.'[14] In another context, he tells us: 'The Tao was the all-inclusive name for this order, an efficacious sumtotal, a reactive neural medium.'[15] I certainly encourage readers to consult Needham's own writings, and very carefully. If you find passages such as these, in their proper contexts to be sure, do not lead you into a terminological roller coaster, you certainly have a better mind than I do.

Moreover, Needham also believes that Einstein's conclusions on the relativistic character of time supports his philosophy. This is revealed by his comment on statements about Chinese ideas of sequential order and retrospective inference. Needham cited Granet:[16]

> The conviction that the universe and each of the wholes composing it have a cylical nature, undergoing alternations, so

> dominated [Chinese] thought that the idea of successesion was always subordinated to that of interdependence. Thus retrospective explanations were not felt to involve any difficulty. Such and such a lord, in his lifetime, was not able to obtain the hegemony, because, after his death, human victims were sacrificed to him.

To this passage Needham commented, 'Both facts were simply part of one timeless pattern.'

In a different context, where Needham discussed a series of ancient Chinese conundrums, we see clearly that his idea of 'timeless pattern', or timeless interdependence, is rooted in his misunderstanding of special relativity. In discussing 'The Paradoxes of Hui Shi', Needham commented on 'going to the State of Yueh, one arrives there yesterday' this way: this sentence[17] 'sounds as if it came out of a modern text-book of physical relativity; it recognized the existence of different time scale in different places.' This is indeed another case of spurious 'anticipation' of modern science — a common problem in the study of the history of science, and a rather uncommon problem when such misunderstanding is used to support an all-embracing philosophy.

I will try to clarify these misunderstandings in the next two sections. Here it may be useful to say a few words to distinguish retrospective inference from determinism, or fatalism. A deterministic course of events is decided by the 'differential equations' that prescribe the 'law of motion', together with the 'initial', or the 'final', or any moment's, conditions, that specify the course. When the above ancient Chinese statement is expressed in a deterministic, instead of retrospectively inferential, fashion, it should read: Such and such a lord, in his lifetime, was not able to obtain the hegemony, because, as one certain event in his life shows, his fate destined him to use human sacrifices after his death.

5 How Does an Average Physicist Understand Physical Causality?

It is true that there was in the nineteenth century the historical development from 'billiard-ball' physics to field physics. But the development was not a replacement of the former by the latter. Rather, it was an extension of 'billiard-ball' physics. Before the advent of the concept of field, people had to accept universal gravity and electromagnetic interactions as mysterious 'actions at a distance'. Now a field of force got rid of this mysteriousness by conceiving fields as physical objects, the propagation of which through space takes time, and which interact with material particles or fields. Therefore, the idea of 'mechanical impulsion' was not only not cancelled, but it extended into the realm of field physics, and asserted that the interactions between fields and particles, and between fields and fields, like that between particles and particles, are all mechanical impulsions by nature.

The innocent clarity of 'billiard-ball' interactions should by no means be used as a reason to look down upon them. Particle physicists have taken millions of photographic records of particle interactions. These records testify without exception that interactions do *occur* according to probabilistic (quantum) laws. They also testify, without exception, that whenever an interaction occurs, elementary particles interact with each other like 'billiard-balls' — although we had better conceive them generally as variable 'billiard-balls' subject to a set of conservation and non-conservation rules. If you ask: Where are the fields? The answer is: Fields are present in the calculation of the probabilities of interactions and their consequent results. But the physical manner of interactions demonstrates, without exception, mechanistic causality. The prediction and experimental confirmation about 50 years ago of the type of elementary particles named meson, can be naturally cited as a theoretical triumph of the 'billiard-ball' concept. Since we are also concerned with the problem of reducibility of life phenomena, it may be useful to add that the particle interactions we are talking about occur at levels many orders beneath the level of the essential biochemical structures and transformations — remember at any moment cosmic rays shower and penetrate copiously every form of life.

After field physics, theoretical physics made progress first in confirming the identity of matter and energy and then in the quantisation of all physical fields. It is obvious that these developments further demonstrate the validity of the 'billiard-ball' concept. Even before the quantisation of fields, Needham's idea of separating connection from causation was incorrect. Are not hydrodynamics, and dynamics of deformable bodies in general, derivative parts of Newtonian particle dynamics? Thus his division between the so-called 'particulate' and 'circumambient' interactions was wrong, because the latter are logically rooted in the former. Thus his assertion of two mutually exclusive 'world views' was wrong, because the very beauty of field physics lies in the logical unity of the two, and in field physics the thorough materialism and clear-cut mathematisation are naturally combined.

Aside from the imperative direct-contact causation, special relativity completed the picture with an important assertion, that is the principle of maximum speed. All matter or energy travels at a speed less than or equal to the speed of light. We are now in a position to formulate the law of physical causality: Any physical causation is effected by matter or energy transmission; the maximum speed of transmission is that of light.

When we say 'any physical causation', we actually understand that nothing is non-physical. To borrow Erwin Schrodinger's phrases, this is a 'naive physicist's' 'comparatively simple and clear and humble' view on causality.

6 Two Contrasting Views on Physical Causality

People generally take a common-sensical stand towards mysticism in causality. In fifteenth century Europe (Switzerland and Germany), archers were often the first candidates as suspects of practising witchcraft. A plausible explanation is that archers are the nearest group of people who could effect an 'action at a distance'. But even a top archer has to use his physical outfits and take physical actions to kill an animal. From our daily experience we all intuitively accept one great principle, namely that a cause precedes its effects, and we reject retrospective inferences.

Now although special relatively made the duration between two events relativistic, namely variable according to different moving observers, it is exactly the principle of maximum speed that prohibits reversing the order of any pair of cause and effect. In this way, Einstein's special relativity provides us a solid theoretical basis for our intuitively established belief. That one left his native state is one event, and that some time later he arrived at another state — another event. The first event was the cause and the second, the effect. Now special relativity asserts that the successive order of these two events can never be changed. This is indeed a great achievement, because thanks to this assertion a common experience of all rational beings is now theoretically confirmed.

As a consequence, we conclude that the ancient Chinese conundrum, 'going to the State Yueh, one arrives there yesterday' is scientifically wrong, and that a textbook of physical relativity is very unlikely to contain the error of stating that this 'Hui Shi paradox' anticipated Einstein in 'recognizing the existence of different time scale in different places'.

More importantly for our present task is that with regard to causality, there is no such thing as 'timeless pattern', or timeless interdependence. Connection, inductance, and resonance, all are effected by travelling agents, matter or energy, hence it takes time, and causes always precede effects. In a field of force, it takes different time intervals for the 'similarly responsive particles' to respond, because they are *pushed* by the propagating field disturbances generally at different times.

In the realm of physiological studies, Needham seemed to be more at ease in talking about the insignificance of the sequential order, or a kind of temporal egalitarianism of events. He commented on a Chinese passage which discusses the Yin and Yang this way:[18]

> The implication was that the universe itself is a vast organism, with now one and now another component taking the lead . . . the larger and the smaller playing their parts according to their degree, 'neither afore nor after other'.
>
> In such a system causality is reticular and hierarchically fluctuating, not particulate and singly catenarian . . . In these phenomena it is not very easy to find out which element is taking the lead at any given time . . . At one time one gland or nerve centre may take the highest place in a hierarchy of causes and effects, at another time another, hence the phrase 'hierarchically fluctuating'.

It is well-known that living creatures grow according to the instructions which are programmed and stored in genetic materials. Unlike Needham, I am not a student of life phenomena, but I would imagine that the daily functioning of living creatures works in about the same manner, that is, taking orders from some sources and carrying these orders out according to environmental conditions. Since so far no biologist has discovered that these orders are transmitted by a mysterious action at a distance, what contradiction is there between 'hierarchically fluctuating' and 'billiard-ball' views of causality — even in the body of a living creature? Nowadays 'taking orders and fulfilling tasks' is a familiar computer practice, and people generally see in this aspect analogies between a living body and a machine. As far as I know, an average computer engineer does not think a computer working by the instruction of its stored program a negation of the mechanistic mode of thought, namely, that information is transmitted by material or energetic agents. When he fathoms the best potentiality of a machine, he does bear in mind special relativity: thc maximum speed of transmission is that of light.

7 Two Contrasting Views on the Reducibility of Biological Phenomena

Because Needham sharply separated 'billiard-ball' physics from field physics, I think a physicist would say that he took a non-physical stand in viewing a physical system. But how does he look at a living system? Certainly once again from a non-physical standpoint, that is a non-reductionist stand.

A reductionist believes that biological phenomena can be made understandable from biochemical changes, and finally from basic physical interactions. By basic physical interactions I mean the transference of matter and energy at a microscopic level, beyond which changes are irrelevant to life phenomena. Since Needham believes such elementary transferences of matter and energy are conducted in a non-physical way, I would regard him as a non-reductionist, or an organicist.

In so far as elementary particle experimentation shows unmistakably the correctness of the law of physical causality at so deep a layer of the material world, I think that the theoretical path to the reductionist or mechanistic goal has already been paved beyond what is necessary, namely the biochemical and physical elucidation of life phenomena. There may always be some 'black box' before the human intellect, like the eternal existence of the mathematical infinitesimal. Or perhaps there will be some final black box, which is ultimately openable. In the latter case, the final physical law that controls the internal mechanism of the final black box may be either deterministic or probabilistic (quantum mechanical). Either way, the future does not influence the present retrospectively: what has been proved is proved; or rather, the 'billiard-ball' interaction, hitherto uniformly shown in all types of records, including visual ones, of elementary particle processes, is unlikely to be disproved one day. Crucial to my argument is that down to a deep enough layer of the material world, direct-contact, or particulate, interactions have been shown to be *the* agency of physical causations. Therefore, in my judgement, neither physics nor biology can be invoked to justify the organicist philosophy. In the case of physics, it is an issue of principle — so long as the law of physical causality holds true; in the case of biology, it is a proven fact, because particulate interactions are the only type of interaction observable beneath the layer of life phenomena. Will some types of interaction operating in between be one day demonstrated that they can never be reduced? If this happens, then my 'proven fact' statement is wrong. But this is a very unlikely prospect. That would mean we live in a 'philosophically uneven' universe: from the microscopic world to the macroscopic world, or reversely, when passing through the *small and thin wafer* of life phenomena, the type of basic interactions changes from 'billiard-ball' to 'endocrine orchestra' (or 'reticulate continuum'), and then back to 'billiard-ball'. (Even were this true, organicist philosophers would not be completely happy, because their mission is to eliminate 'billiard-ball' interactions even from collisions between two billiard balls.) Empirically this is also very unlikely. (That is why I say a 'proven fact'.) Because all the layers of the material world around and within the *small and thin wafer* of life phenomena are at present within the reach of modern sciences, and experimentation on physical interactions at all these attainable layers is being conducted everywhere everyday. What is the clearest evidence that shows some types of interactions are definitely (or even suggestively) irreducible? I am very very curious to know.

8 Quantum Mechanics and Physical Causality

The advent of quantum physics in the 1920s further sensitised the problem of physical causality. The famous 1935 debate between Einstein-Podolsky-Rosen and Niels Bohr[19] on the question of the completeness of the wave-function description of reality was in fact the result of Einstein's longtime suspicion about the statistical nature of quantum mechanics.[20] The gist of many thought-experiments, which were designed by Einstein, Bohr and others to examine the 'dice-cast' character of the new type of physical laws, was to see whether the violation of Heisenberg's Uncertainty

Principle would lead to the violation of the law of physical causality. For us, the important thing is that had Einstein demonstrated that the validity of the Uncertainty Principle destroys the validity of the law of physical causality, the majority of physicists would flock to Einstein's side and abandon quantum mechanics. Fortunately, Einstein did not succeed in this mission, and the happy denouement of the joint efforts of Einstein-Podolsky-Rosen and Bohr (and many others) was: so far there is no contradiction between the Uncertainty Principle (quantum mechanics in general), and the law of physical causality (special relativity in general). In short, in any type of quantum field, 'particulate' or direct-contact causality operates. (We have mentioned above that elementary particles — for which quantum effects are most prominent — interact with each other like 'billiard-balls'.)

Nevertheless, from general reading some people do acquire the impression that the new mechanics implies an essentially different causality. This was to a large extent a result of the use of terminology. Newtonian mechanics predicts the resultant positions and momenta of particles deterministically, whereas quantum mechanics does it statistically. Many writers refer to the indeterministic character of quantum mechanics as being 'non-causal'.[21] Actually this last adjective means statistical, or probabilistic. Classical statistical mechanics is probabilistic, yet it is a 'billiard-ball' physics as well. No one calls classical statistical mechanics 'non-causal'. Now the basic law of quantum mechanics is by nature probabilistic, and the modifier 'non-causal' is used by some writers to stress this fact. For us, the crux is that quantum mechanics conforms to the law of physical causality — the old 'billiard-ball' physics plus the principle of maximum speed. Today very few physicists harbour serious suspicion about the law of physical causality.[22]

9 'Synthesis over Analysis' or 'Analysis Leads Synthesis'?

There is yet another aspect of the problem, namely the methodological aspect. That it is a valid aspect is underlined by, for instance, the fact that Needham is inclined to 'synthesis over analysis' — as Nakayama told us.

I believe that the breakthrough from vague conjectures about the universe as a whole, or about the human body as a whole, to analytic investigations of limited parts of these wholes, the breakthrough from unspecified, ill-defined philosophical inferences to the formulations of 'mechanistic' laws and contact interaction causality, and the breakthrough from using analogy as a dependable means of reasoning to exact and step-by-step logic, represent a decisive stage towards the growth of modern science. Every mathematical operation, from simple arithmetical ones to integrating a differential equation, or to summing up the distribution of prime numbers, is based upon digit-to-digit and point-to-point 'movement'. A mathematical operation may look grandiose in total, still it is built upon one-by-one counting, or continuous moving. In brief, mathematical synthesis follows mathematical analysis. Biological approaches, like the progress of microscopic physics, appear to proceed in the opposite direction. From the simple observation 'like begets like' to the Mendelian law, and to the discovery of the genetic double helix, black boxes have been opened one after another, pointing to the same philosophical essence: internal mechanisms exist, Man is entitled to find them out, and finally to ascertain the role of every next atom in a DNA molecule.

After nearly 400 years, the mainstream of modern science still develops along this analytic and 'mechanistic' guideline, a guideline which consistently seeks physical, direct-contact causalities. So far I see no sign of serious challenges to this tendency, still less of its replacement or reorientation. Methodologically, not mechanically (isn't it ironic?), analysis must precede synthesis. Neither incompetence by the human intellect in analysis, nor a personal 'keen propensity for synthesis', comprises a valid excuse to change this general rule of epistemology. My short critique amounts to a defence of mechanistic causality against Needham's challenge, and an effort to reduce the 'organicist haziness'.

In the final analysis, the organicist philosophy accepts some vague association between a cause and an effect as final. To use Nakayama's general interpretation:[23] 'Things to be grasped in terms of impact in the West are comprehended as resonance by the Chinese, and hence the latter can conceive action-at-distance more easily.' This correlative philosophy asserts either that further analysis is impossible or that more intimate correlations between a cause and an effect simply do not exist. (In order to reach organicist conclusions such as these, the necessary methodology is analysis. Therefore I really do not understand what it means by 'synthesis over analysis'.) As I have said before (note 5 of the Introduction), the former view may represent a theoretical reductionist and practical non-reductionist. Thus the orthodox organicist ground is very limited and, thanks to the rapid progress in scientific experimentations in all attainable material levels — none of which is not related, one way or another, to the essence of physical interactions — I also believe that this gound is rapidly wearing thin.

To summarise we can make the following contrast, as set out in Table 4.1.

Table 4.1: Two Contrasting Scientific Philosophies

Two Philosophies	Physical	Organicist
on physical causality	direct contact, or mechanistic view	'field of force', or 'action at a distance', view
on the reducibility of life phenomena	reducible — a reductionist view	irreducible — a non-reductionist view
on basic methodology	analysis leads	synthesis dominates

(Here 'field of force' is put in quotation marks because, as we have made clear above, it is a misunderstanding on the part of organicist philosophers; 'action at a distance' is put in quotation marks because it has been a convention — it has been an outdated misnomer since the introduction of the concept of a field of force.)

Finally a legitimate question arises: is it possible for one to hold a mixed point of view on physical causality and reducibility? If one believes that any causation is, in the final analysis, effected by elementary transference of matter and energy, and that life phenomena are no exception to this rule, he is a reductionist. On the other hand, a person who holds an 'action at a distance' view on physical causation is liable to profess whether he believes life phenomena are reducible to his physical 'actions at a distance', or to other types of 'actions at a distance', which are peculiar in living bodies. If he belongs to the former group, he is a reductionist biologist, but a non-mechanical physicist, or a non-physical physicist. If he subscribes to the latter belief, he is a true, or philosophical, organicist. From what we have read, Needham appears to be a true organicist, or an 'organic philosopher'.

10 Some Concluding Remarks

Finally, there was a mathematical incident, which was also thought to support the 'Chinese correlativism'. Needham believed that geometry was characteristic of Greek mathematics, and algebra was characteristic of Chinese mathematics.[24] Delightfully he told his readers that 'Whitehead has termed algebra the mathematical study of pattern',[25] and believed this was no coincidence.[26] Nevertheless, to a mathematician (not necessarily as outstanding as Whitehead), it is clear that geometry and algebra are one, though the former is narrower — for among other things, it is limited by the three-dimensional space. Both are 'mathematical study of pattern'. (Historically, hence more intuitively, there is no question that geometry is more pattern-oriented. So we don't know what supports Needham's correlativism, especially here.) In this computer age we may use a modern illustration, namely the analog versus the digital representations of calculated results. While the accuracy of analogous representation is much limited, that of the digital is theoretically limitless. So is geometry to algebra — analogously. Any pattern can be represented algebraically, whereas geometry becomes helpless for the great majority of patterns which are pictorially impossible.

In summary I hope I have in this short critique made clear the following points about Needham's thought: (1) Needham believes that the guiding scientific philosophy and methodology are going through an important transition, from mechanistic to organicist; (2) he thinks highly of some Chinese proto-scientific, rather than proto-philosophic, thought; and (3) his logical linkage of these two beliefs. There is another related Needhamite speculation, which is historical, hard to confirm or to deny. That, as I have mentioned in Section 4, Essay One (Chapter 2) and Section 15, Essay Two (Chapter 3), is about the probable Chinese origin of the Leibnizan organicist thought.

The physical and scientific aspect of my critique is to point out that organicist philosophy is very unlikely to guide our future science and scholarship in general. Billiard-ball physics (which unfortunately has been ridiculed time and again by Needham) and field physics are not mutually exclusive,[27] and field physics does not support an unspecified organicist view on natural systems. Neither does 'endocrine orchestra', or 'reticulate continuum', biology. My primary concern was China's long stagnation, or inertia. Therefore, naturally, it was Needham's glorifying appraisal of Chinese intellectual achievements, including scientific philosophy, that engendered suspicion in my mind. I do think that we should not give undue credits to China's present as well as past. Because China has been evidently an ailing giant, suffering from a persistent, chronic disease.

Needham endeavours to demonstrate the ancient glory of Chinese scientific, philosophical, and technological traditions, in order to correct a 'total misconception' about China. However, from Needham's second volume, I want to requote a passage, which may represent a typical 'Western misconception', to support my own view on Chinese development. Levy-Bruhl wrote:[48]

> Chinese scientific development affords a striking example of this arrested development. It has produced immense encyclopaedias of astronomy, physics . . . and yet to our minds all this is nothing but balderdash. How can so much effort and skill have been expended in the long course of ages, and yet their product be absolute nil? This is due, no doubt, to a variety of causes, but above all to the fact that the foundation of each of these so-called sciences rests upon crystallised concepts, concepts which have never really been submitted to the test of experience, and which contain scarcely anything beyond vague and unverifiable notions with mystical preconnections . . . Those who are best acquainted with the Chinese mentality . . . almost despair of seeing it free itself from its shackles, and cease revolving on its own axis . . .

In the main, I agree with this passage on a descriptive level, and I identify myself with those who despair of seeing China cease to revolve on its own axis. The idea that the pre-modern Chinese mode of thinking took no interest in testing concepts by experience reminds us what David Joravsky described as 'some dreamy quality' in traditional mentality which permitted persistent, false, and easily disprovable beliefs (Section 4, Essay One, Chapter 2). Yet I must add a qualification to Levy-Bruhl's passage to balance the record: that Western Europe finally produced modern science was, in a sense, because it had a much bigger, and more importantly, more variegated store of 'balderdash'. About this very passage Needham remarks, 'It would be hard to find a passage more misguided.'

I don't know if my readers have acquired the impression that Needham loves China more than I do. But I can assure you that would be a false impression; it is no more true than that I speak better English than does Needham.

Notes

1. S. Nakayama, 'Joseph Needham, Organic Philosopher' in S. Nakayama & N. Sivin (eds), *Chinese Science* (The MIT Press, Cambridge, Mass., 1973), p. 25.
2. Ibid., pp. 25–6.
3. Ibid., p. 39.
4. J. Needham, *Science and Civilization in China*, vol. II (Cambridge University Press, Cambridge, Mass., 1956), p. 339.
5. Ibid., p. 288.
6. Ibid., p. 340.
7. Nakayama, note 1 above, p. 39.
8. Needham, note 4 above, pp. 280–1.
9. Ibid., p. 281.
10. Ibid., p. 339.
11. Ibid., p. 291.
12. Ibid., p. 285.
13. Ibid., p. 289.
14. Ibid., p. 293.
15. Ibid., p. 290.
16. Ibid., p. 289.
17. Ibid., p. 193.
18. Ibid., p. 289.
19. A. Einstein, B. Podolsky and N. Rosen, 'Can Quantum Mechanical Description of Physical Reality Be Considered Complete?' *Physical Review*, vol. 47 (May 1935), pp. 777–80; N. Bohr, 'Can Quantum Mechanical Description of Physical Reality Be Considered Complete?' *Physical Review*, vol. 48 (Oct. 1935), pp. 696–702.
20. N. Bohr, 'Discussion with Einstein on Epistemological Problems in Atomic Physics' in P. Schilpp (ed.), *Albert Einstein: Philosopher-Scientist* (The Library of Living Philosopher, Inc., 1949).
21. See, e.g., Max Boorn, 'Einstein's Statistical Theory' in *Albert Einstein: Philosopher-Scientist* (note 20 above).
22. After reviewing my manuscript, Dr Robert Michaelson kindly commented on this question and called my attention to the fact that some physicists do share the philosophy of organism. The following two paragraphs are from Dr Michaelson:

> Although I agree in general with you, I must point out that one can usually find a distinguished modern physicist who supports virtually any philosophical view one chooses. I suspect that G. Chew or B. Josephson might well support some of Needham's view. As a specific example, Pauli took very seriously the ideas of the psychoanalyst Jung on 'archetypes', 'synchronicity', etc. To quote from Pauli (English anstract of '. . . Aspekt der Ideea von Unbewussten' in *Collected Scientific Papers*, vol. 2, p. 1239):
> ' "correspondence", "complementary pairs of opposites" and "wholesomeness" appear independently both in physics as well as in the ideas of the unconscious. The "unconscious" itself has a certain analogy to the "field" in physics and

> both are shifted by an observational problem, outside the range of visuability into the paradoxical. Although in physics one does not speak of "archetypes" which reproduce themselves, but of "statistical laws of nature with primary probabilities", both formulations meet in the tendency to amplify the older more narrow idea of "causality" (determinism) to a more general form of "connections" in nature, toward which the psychophysical problem also points.'

Although these two paragraphs serve my paper as a welcome balance, I certainly cannot be easily convinced. In order to champion my stand I want to point out that Jung is generally known as a mystic.

23. Nakayama, note 1 above, p. 39.
24. Needham, note 4 above, p. 292.
25. Ibid.
26. Ibid.
27. I am aware that the 'billiard-ball–field physics problem' can be discussed from historical, philosophical, quantum-theoretic, and experimental perspectives respectively, and all at great length. In this chapter, however, I stress only either 'billiard-ball' physics as an experimental reality down to the very depths of the material world, or 'billiard-ball' physics as the conceptual basis of field physics. Since both these two aspects are crystal clear to every physicist, there is no need to have recourse to prolixity. Therefore, I believe the discussion, though short, is sufficient for our purpose, namely, to clarify some confounded ideas with regard to the scientific justification of scientific philosophies and methodologies.
28. Needham, note 4 above, p. 285.

INDEX